PLANT SCIENCE RESEARCH AND PRACTICES

NIGELLA SATIVA

PROPERTIES, USES AND EFFECTS

PLANT SCIENCE RESEARCH AND PRACTICES

Additional books and e-books in this series can be found on Nova's website under the Series tab.

PLANT SCIENCE RESEARCH AND PRACTICES

NIGELLA SATIVA

PROPERTIES, USES AND EFFECTS

SANJIN BERGHUIS
EDITOR

NOTICE TO THE READER

Library of Congress Cataloging-in-Publication Data

ISBN: 978-1-53617-538-7
Library of Congress Control Number: 2020932281

Published by Nova Science Publishers, Inc. † New York

CONTENTS

PREFACE

Nigella sativa: Properties, Uses and Effects outlines current knowledge on the nutritive value of *Nigella sativa* (black cumin). *N. sativa* has many therapeutic effects and is considered one of the most important medicinal plants in the world because of its antioxidant, anticoccidial, anthelminthic and antimicrobial properties.

Based on the use of *N. sativa* in traditional medicine as a treatment for some diseases, researchers have investigated its effects on asthma, hypertension, diabetes, and inflammation.

Much of the biological properties of *N. sativa* including anti-hypertensive nephroprotection, antipyretic, antimicrobial, and antineoplastic has been attributed to presence of thymoquinone. As such, the authors accounts for therapeutic potential of thymoquinone.

Additionally, a field experiment was conducted to determine the influence of plant density and fertilization on seed and biomass yield and quality of *N. sativa* in order to define alternatives to local forage and feed sources for animals feeding in the Mediterranean region.

Chapter 1 - This review outlines the current state of knowledge on the nutritive value of black cumin (*Nigella sativa* L.). The popularity of this plant, which is an annual herbaceous plant belonging to the *Ranunculaceae* family and is native to Iran, Pakistan and Turkey, is due to its beneficial actions. Black cumin has many therapeutic effects and is considered one of the most important medicinal plants in the world because of its antioxidant, anticoccidial, anthelminthic and antimicrobial activities. The nutritive value of black cumin is result of its carbohydrate, fatty acid, protein contents as well as its several bioactive compounds. The seeds or their byproducts can be used in feeds for farm animals, with positive effects on the compositional characteristics of eggs, milk and meat.

Chapter 2 - Black cumin (*Nigella sativa* L.) belongs to Ranunculaceae family. As an aromatic plant, *N. sativa* is widely grown in different parts of the world and the seeds of black cumin have been used to promote health for Middle East and Southeast Asia countries. *N. sativa* seeds yield esters of fatty acids, free sterols and steryl esters. The seeds also contain lipase, phytosterols and sitosterol. During the past few decades, many phytochemical and pharmacological studies have been conducted on *N. sativa* seeds because of its marked biological activities, antioxidant, anti-inflammatory and antiulcer activity. The main compounds that have been isolated from *N. sativa* seed include proteins, carbohydrates, fixed oils, essential oil, crude fiber, alkaloids, minerals, vitamins, ash, and moisture. The other components are tannins, resin, saponin, carotene, glucosides and sterols. One of the major components of the essential oil is thymoquinone (TQ). Based on the use of *N. sativa* in traditional medicine as a natural treatment for some diseases, researchers have investigated its protective effects against asthma, hypertension, diabetes, and inflammation. In addition, TQ is known to have antioxidant, antifungal, antibacterial, anticancer and neuroprotective activities. It has been found that TQ has beneficial protective effects against renal diseases through anti-inflammatory, antioxidant, and antiapoptotic activities. Here, the authors are giving a summary of antioxidant, anticancer, antimicrobial effects of *N. sativa* on regard of pharmacological uses.

Chapter 3 - Recently the World Health Organization has urged and encouraged countries of the developing world to upgrade their primary healthcare programs with the incorporation of traditional medicinal plant therapy. One of the most studied medicinal plants is *Nigella sativa*. *N. sativa* from the family *Ranunculaceae* is an annual flowering plant also called black cumin, black seed, or Habbatul Barakah. It is also described as the "miracle herb" of the century. *N. sativa* is native to south and southwest Asia wherein the plant is cultivated and grows. It is also widely cultivated in Mediterranean countries, middle Europe and western Asia. It has amazing curative and therapeutic features that make them one of the most popular, safe, non-detrimental, and cytoprotective medicinal plant that can be used for prevention and treatment of many complicated diseases. *N. sativa* seeds and oil extract have been known to be positively implicated in allergy, diabetes, cardiovascular problems like hypertension, hyperlipidemia, gastro- intestinal problems, inflammatory and oxidative damage processes. Moreover, its antihistaminic, antioxidative and immunomodulator properties cover a wide range of degenerative health problems apart from few recent reports of its anti-viral properties. *N. sativa* seeds contain diverse but well-characterized chemical components; which include both fixed and essential (volatile) oil, proteins and amino acids, carbohydrates, alkaloids, organic acids, saponins, crude fibers, vitamins, and minerals. Thymoquinone (2-isopropyl-5-methylbenzo-1, 4-quinone) (TQ), the most abundant constituent of the volatile oil also present in the fixed oil, is the biologically active component of *N.sativa* seeds. TQ as a bioactive component is found in many medicinal plants. Apart from *Ranunculaceae* family, the presence of this compound has been confirmed in several genera of the *Lamiaceae* family such as *Agastache*, *Coridothymus*, *Thymus*. It has also been found in genus *Teraclinis* and in the form of glycoside in Juniperus of *Cupressaceae* family. *N.sativa* seeds contain more than 100 chemical compounds, a number of them are yet to be characterized. However, much of the biological activity of the *N. sativa* including antihypertensive nephroprotection, antipyretic, antimicrobial, and antineoplastic has been attributed to presence of TQ in it. The major pharmacological activities demonstrated by TQ are anticonvulsant, antimicrobial, anticancer, anti-histaminic, antidiabetic, anti-inflammatory, and antioxidant. More recently, a great deal of attention has been given to this dietary phytochemical with an increasing interest to assess its health benefits through pre-clinical and clinical researches. Considering its wide range of medical applications TQ can be a potent candidate to be used as a natural drug. Considering the extraordinary medicinal attributes of TQ, this chapter accounts for therapeutic potentials of TQ.

Chapter 4 - *Nigella sativa* L. has been recognized as one of the most important medicinal plants in many parts of the world for centuries. It is also a valuable source of carbohydrates, proteins, essential fatty acids, vitamins, and minerals. Because of its exceptional nutritional and medicinal properties, this plant has been considered as one of the main sources of nutrition and healthcare for humans and animals. Literature survey revealed that dietary *N. sativa* seeds greatly improved animal products quality without compromising growth, productivity, as well as organoleptic quality. *N. sativa* is mainly cultivated as a seed crop, and there is no information on biomass quality and its potential for animal feeding. A field experiment was conducted to determine the influence of plant density and fertilization on seed and biomass yield and quality of *N. sativa* crop in order to define alternatives to local forage and feed sources for animals feeding in the Mediterranean region. The 2-year experiment was laid out in a split-plot design with two main plots (plant densities: 200 and 300 plants m^{-2}), four sub-plots [fertilization treatments: control (untreated), seaweed compost, farmyard manure, and inorganic fertilizer] and three replications for each treatment. The results indicated that *N. sativa* yields were affected by both plant density and fertilization. The highest seed yield (749-800 kg ha^{-1}) was found in low density plots fertilized with inorganic fertilizer, while the highest biomass yield (3755-4346 kg ha^{-1}) was observed in high density plots fertilized with inorganic fertilizer. Concerning the crude protein (CP) content, the highest content of total aboveground biomass (22.43-24.17%) was measured at 75 days after sowing and was achieved in plots with low density and inorganic fertilization. The highest seed CP content (26.94-27.45%) was also found under low density and inorganic fertilization. Ash and ether extract content were not influenced by plant densities and fertilization regimes. In addition, acid detergent fiber (ADF) and neutral detergent fiber (NDF) were not affected by plant density but there were significant differences between fertilization treatments. The highest ADF (35.5-36.8% and 19.7-20.0% for total aboveground biomass and seed, respectively) and NDF

(49.2-50.1% and 30.4-31.1% for total aboveground biomass and seed, respectively) were found under inorganic fertilization at 115 DAS. As a conclusion, *N. sativa* could be successfully used as a novel feed crop.

In: *Nigella sativa*
Editor: Sanjin Berghuis
ISBN: 978-1-53617-538-7

Chapter 1

NUTRITIVE VALUE OF BLACK CUMIN (*NIGELLA SATIVA* L.)

Pier Giorgio Peiretti*
Institute of Sciences of Food Production,
National Research Council, Grugliasco, Italy

ABSTRACT

This review outlines the current state of knowledge on the nutritive value of black cumin (*Nigella sativa* L.). The popularity of this plant, which is an annual herbaceous plant belonging to the *Ranunculaceae* family and is native to Iran, Pakistan and Turkey, is due to its beneficial actions. Black cumin has many therapeutic effects and is considered one of the most important medicinal plants in the world because of its antioxidant, anticoccidial, anthelminthic and antimicrobial activities. The nutritive value of black cumin is result of its carbohydrate, fatty acid, protein contents as well as its several bioactive compounds. The seeds or their byproducts can be used in feeds for farm animals, with positive effects on the compositional characteristics of eggs, milk and meat.

Keywords: black cumin, fatty acid, protein, carbohydrates, byproducts

* Corresponding Author's E-mail: piergiorgio.peiretti@ispa.cnr.it.

INTRODUCTION

Thousands of research articles that have been published in international medical journals are available on the internet regarding studies on the medicinal properties the seeds of black cumin (BC, *Nigella sativa* L.), their oil and their bioactive compounds (Randhawa, 2008). Over the last two decades, several reviews have provided comprehensive information on the nutritional, health, pharmaceutical, therapeutic and biomedical applications and prospects of BC (Randhawa and Al-Ghamdi, 2002; Ali and Blunden, 2003; Gilani et al., 2004; Salem, 2005; El-Tahir et al., 2006; Gali-Muhtasib et al., 2006; Akhondian et al., 2007; Ramadan, 2007; Ismail and Yaheya, 2009; Sharma et al., 2009; Paarakh, 2010; Randhawa and Alghamdi, 2011; Naz, 2011; Shrivastava et al., 2011; Mirzaei, 2012; Datta et al., 2012; Ahmad et al., 2013; Alenzi et al., 2013; Azeem et al., 2014; Forouzanfar et al., 2014; Mollazadeh and Hosseinzadeh, 2014; Razavi and Hosseinzadeh, 2014; Tembhurne et al., 2014; Aljabre et al., 2015; Bamosa, 2015; Gholamnezhad et al., 2015; Heshmati and Namazi, 2015; Longato et al., 2015; Mahdavi et al., 2015; Majdalawieh and Fayyad, 2015; Amin and Hosseinzadeh, 2016; Beheshti et al., 2016; Gholamnezhad et al., 2016; Hayatdavoudi et al., 2016; Hussain and Hussain, 2016; Khan et al., 2016; Kooti et al., 2016; Mohtashami and Entezari, 2016; Sahebkar et al., 2016a, 2016b; Shokri, 2016; Ijaz et al., 2017; Tavakkoli et al., 2017; Mazaheri et al., 2019).

Several studies performed on experimental animals have also demonstrated the principal effects of BC seeds or their bioactive compounds (El-Komey, 1996; Nagi et al., 1999; Mansour et al., 2001; Bamosa et al., 2002; Kanter et al., 2003, 2005; Abdel-Wahhab and Aly, 2005; Al-Othman et al., 2006; Hosseinzadeh et al., 2007; Perveen et al., 2008; Pourghassem-Gargari et al., 2009; Kanter, 2009; Khaled, 2009; Ismail et al., 2010; Harzallah et al., 2012; Umar et al., 2012; Ahmad and Qasem 2017).

Besides being a significant source of lipids, proteins and carbohydrates, BC also contains over 100 bioactive compounds, such as carotene, essential fatty acids, galactolipids, minerals, phenolic compounds, phospholipids, sterols and vitamins (Takruri and Dameh, 1998). Because of this interesting composition, BC has various therapeutic effects and is considered one of the most important medicinal plants in the world due to its antioxidant, antiinflammatory, anticancerous, anticoccidial, anthelminthic and antimicrobial activities (Atta and Imaizumi, 1998; Burits and Bucar, 2000; Morsi, 2000; Schwarz et al., 2001; Maqbool et al., 2004; Aljabre et al., 2005; Rahman and Nada, 2006; El-Gendy et al., 2007; El-Shenawy et al., 2008; Erkan et al., 2008; Baghdadi and Al-Mathal, 2011; Chaieb et al., 2011; Bourgou et al., 2012; Harzallah et al., 2012; Pichette et al., 2012; Petrujkić et al., 2018).

Furthermore, several studies have dealt with the chemical composition and nutritional aspects of black cumin, and this review outlines the present state of knowledge on the nutritive value of BC.

NUTRITIVE VALUE

The chemical composition, as determined by different authors, of BC seeds or BC cake is reported in Table 1.

Table 1. Percentage composition of black cumin (*Nigella sativa* L.) on a dry matter basis

	Al-Jassir (1992)	**Al-Gaby (1998)**	**Takruri & Dameh (1998)**	**Awadalla & Gehad (2003)**	**Atta (2003)**
Product	seed	cake	seed	seed	seed
Dry matter	95.4	-	96.2	95.4	93.0
Crude protein	20.9	16.1	21.6	25.5	20.8
Carbohydrate	31.9	38.4	24.9	17.1	33.7
Ether extract	38.2	31.4	40.6	43.3	34.8
Crude fiber	7.9	7.1	8.4	8.5	-
Ash	4.4	5.5	4.5	5.5	3.7
Gross energy (MJ/kg)	-	-	23.1	-	-

Table 2. Amino acid composition (g/100 g protein) of black cumin (*Nigella sativa* L.)

	Al-Jassir (1992)	Al-Gaby (1998)	Mariod et al., (2012)
Product	seed	cake	seed
Aspartic acid	8.9	10.0	15.8-16.0
Threonine	3.6	3.9	7.7-8.4
Serine	4.3	3.8	5.5-6.9
Glutamic acid	24.7	22.4	41.7-43.2
Glycine	5.6	6.9	10.2-11.2
Alanine	3.7	4.2	9.0-9.6
Cystine	2.0	1.2	3.6-37
Valine	4.6	5.1	12.0-12.7
Methionine	1.6	1.4	2.1-2.7
Isoleucine	3.5	4.0	9.0-9.6
Leucine	5.8	6.9	13.0-13.7
Tyrosine	3.6	3.3	3.0-4.0
Phenylalanine	3.6	4.0	6.9-7.6
Histidine	3.3	2.8	5.5
Lysine	4.0	3.9	8.8-9.3
Arginine	9.2	9.2	18.3-19.9
Proline	4.9	6.1	10.7-11.3
Tryptophan	n.d.[a]	0.77	n.d.

[a] n.d. = not determined.

As far as the amino acid composition is concerned, lysine, cysteine and methionine, that are usually limiting in plant foods, are present in good quantity in BC (Takruri and Dameh, 1998). Various authors have found that the most abundant amino acids in BC are: glutamic acid, arginine, aspartic acid, leucine, glycine, proline and valine, as reported in Table 2.

The variation in the chemical composition and in the amino acid content could be due to the differences in the analytical techniques that were used to establish the contents or may even be attributed to varietal or geographic and/or climatic differences in the areas where the plants were grown.

Al-Gaby (1998) found that BC cake protein can be used as a complementary agent in both corn meal and bean meal protein without any adverse nutritional effects concerning the lipid fractions levels in the serum of experimental diet-fed rats.

Table 3. Fatty acid profile (g/100 g total fatty acid) of black cumin (*Nigella sativa* L.) fixed oil

	Ramadan & Mörsel (2003)	Atta (2003)	Nickavar et al., (2003)	Cheikh-Rouhou et al., (2007)	Hamrouni-Sellami et al., (2008)
C12:0	n.d.[a]	n.d.	0.6	n.d.	n.d.
C14:0	n.d.	9.8-11.1	0.5	0.4	3.2
C14:1	n.d.	tr[b]	n.d.	n.d.	n.d.
C16:0	16.9-20.1	9.9-12.1	12.5	17.2-18.4	12.2
C16:1	n.d.	0.5-0.7	n.d.	0.8-1.2	n.d.
C18:0	1.6-2.3	3.3-3.7	3.4	2.8-3.7	6.3
C18:1	22.9-27.2	18.9-20.1	23.4	23.7-25.0	12.7
C18:2	48.5-52.8	47.5-49.0	55.6	49.2-50.3	61.3
C18:3	n.d.	2.1-2.7	0.4	0.3	1.5
C20:0	n.d.	0.7-1.2	n.d.	0.1-0.2	0.2
C20:1	0.4-0.7	n.d.	n.d.	0.3	0.4
C20:2	2.0-2.4	n.d.	n.d.	n.d.	n.d.
C22:0	n.d.	0.8-0.9	n.d.	2.0-2.6	2.2
C22:1	n.d.	0.7-1.0	n.d.	n.d.	n.d.
C24:0	n.d.	0.2-0.3	n.d.	tr	n.d.

[a] n.d. = not determined.

[b] tr = traces.

The BC seed oil profile reported in the literature is shown in Table 3 with reference to the fatty acid (FA) content. A higher amount of unsaturated FAs is nutritionally desirable, as it has certain positive effects on animals supplemented with this oil. Ramadan and Mörsel (2003) found six FAs, with the dominating FA being linoleic acid (C18:2n6), followed by oleic (C18:1n9) and palmitic acid (C16:0). Atta (2003) found that the major unsaturated FAs in BC seed oil were oleic, linoleic and linolenic acid (C18:3n3), while palmitic, myristic (C14:0) and stearic acid (C18:0) were the main saturated FAs; moreover, there were minor amounts of arachidic (C20:0), behenic (C22:0), lignoceric (C24:0), palmitoleic (C16:1n7) and erucic acid (C22:1n9). Cheikh-Rouhou et al., (2007) and Hamrouni-Sellami et al., (2008) reported a negligible amount of eicosenoic acid (C20:1). Some minor FAs had not been detected in previously published data (Üstun et al., 1990; Al-Jassir, 1992; Abdel-Aal and Attia, 1993).

The variations in FA composition may be due to different causes: the seed quality (maturity, harvesting-caused damage and handling/storage conditions), environmental, genetic (variety grown, plant cultivar), oil-processing variables or the accuracy of detection and quantitative techniques (Ramadan and Mörsel, 2002). Salem (2001) studied the effect of heat treatments on some physical and chemical properties of BC seed oil.

Table 4. Black cumin (*Nigella sativa* L.) minerals (mg/kg)

	Takruri & Dameh (1998)	Özcan (2004)	Ashraf et al., (2006)	Mariod et al., (2012)
Potassium	5257	6932	6750	426-589
Phosphorus	5265	4504	3781	n.d.[a]
Sodium	496	n.d.	4125	226-368
Calcium	1859	9062	2809	17-19
Iron	105	181	104	3.5-16
Zinc	60	50	44	3.5-5.2
Magnesium	n.d.	3788	1382	n.d.
Manganese	n.d.	33	125	0.99-1.5
Copper	18	11	32	0.83-0.94
Nichel	n.d.	n.d.	91	n.d.

[a] n.d. = not determined.

Different inorganic elements such as K, P, Na, Ca, Fe, Zn, Mg, Mn, Cu and Ni, have been found in BC seeds (Table 4). Takruri & Dameh (1998) reported that four minerals (P, Cu, Fe and Zn) were relatively high in BC and concluded that the nutritional status of these minerals cannot be predicted from the apparent quantity. Özcan (2004) determined the mineral contents of several plants and found that the mineral elements varied greatly, depending on the species. Ashraf et al., (2006) assessed the pattern of accumulation of some of these inorganic elements in the seeds of BC plants grown at various Nitrogen levels and reported that the K, P, Ca and Mg contents were not affected as the N application rate was increased, while a consistent decrease in seeds was observed for Mn, Zn and Ni for an increasing N rate. Mariod et al., (2012) studied the effect of different germination conditions on the contents of two BC seeds and found that germinated seeds showed noticeable decreases in the K, Na, Ca and Fe contents, due to the leaching of such minerals into the soaking water.

As far as the vitamin content of BC seeds is concerned, Takruri and Dameh (1998) found four vitamins: thiamin (15.4 mg/kg), niacin (57 mg/kg), pyridoxine (5.0 mg/kg) and folic acid (160 µg/kg).

BC seeds are a rich source of several phenolic compounds that appear to have a particularly positive effect on human and animal health. The main constituents of BC seeds are alkaloids, and fixed and volatile oils. Various authors have found extremely different essential oil compositions of BC seeds, as shown in Table 5.

Nickavar et al., (2003) identified thirty-two compounds of which six were phenyl propanoid compounds, nine were monoterpenoid hydrocarbons, four monoterpenoid ketones, eight nonterpenoid hydrocarbons, three monoterpenoid alcohols and two sesquiterpenoid hydrocarbons. The most abundant compounds were trans-anethole (38.3%), p-cymene (14.8%), limonene (4.3%) and carvone (4.0%).

Moretti et al., (2004) identified nineteen compounds, and reported that BC essential oil was characterized by high p-cymene (33.8%) and thymol (26.8%) contents, and these were followed by iso-3-thujanol (7.4%), β-elemene (5.5%), thymoquinone (3.8%), α-thujene (3.3%) and longifolene (3.1%).

Singh et al., (2005) identified thirty-eight components, and found that the major component was p-cymene (36.2%), followed by thymoquinone (11.3%), α-thujene (10.0%), longifolene (6.3%) and carvacrol (2.1%).

Hamrouni et al., (2008) found eighteen volatile compounds that belonged to four chemical classes: monoterpenic hydrocarbons (86.3% of the total volatiles) with p-cymene (53.1%) and ocimene (18.5%) as the major compounds; monoterpenic alcohols (7.1% of the total volatiles) with octen-3-ol (6.5%) as the major compound; terpenic ethers, which were only represented by 1,8-cineole (1.9%); and monoterpenic phenols, which were only represented by thymol (1.8%).

Table 5. Percentage composition of black cumin (*Nigella sativa* L.) essential oil

	Nickavar et al., (2003)	Moretti et al., (2004)	Singh et al., (2005)	Hamrouni et al., (2008)	Wajs et al., (2008)
α-Thujene	2.4	3.3	10.0	7.2	7.2
α-Pinene	1.2	0.7	3.3	1.4	2.0
β-Pinene	1.3	1.1	3.8	1.8	2.1
Sabinene	1.4	0.5	1.3	0.7	0.8
p-Cymene	14.8	33.8	36.2	53.1	60.2
2,4(10)-Thujadiene	n.d.[a]	n.d.	0.1	n.d.	n.d.
Camphene	n.d.	n.d.	0.1	n.d.	tr [b]
cis-Sabinene	n.d.	n.d.	0.2	n.d.	n.d.
trans-Sabinene	n.d.	1.0	0.2	n.d.	n.d.
α-Phellandrene	0.6	n.d.	n.d.	0.1	0.2
β-Myrcene	0.4	0.3	n.d.	2.1	0.4
o-Cymene	n.d.	3.3	n.d.	18.5	tr
α-Terpinene	n.d.	0.6	0.5	n.d.	tr
γ-Terpinene	0.5	2.4	0.2	1.2	12.9
Limonene	4.3	1.1	1.8	0.1	1.3
Terpinolene	n.d.	n.d.	0.1	0.1	0.6
cis-Thujan-4-ol	n.d.	n.d.	n.d.	n.d.	tr
trans-Thujan-4-ol	n.d.	n.d.	n.d.	n.d.	0.5
cis-4-Methoxythujane	n.d.	n.d.	n.d.	n.d.	tr
β-Thujone	n.d.	n.d.	n.d.	n.d.	tr
trans-4-Methoxythujane	n.d.	n.d.	n.d.	n.d.	4.0
iso-3-Thujanol	n.d.	7.4	n.d.	n.d.	n.d.
Camphor	n.d.	n.d.	0.1	n.d.	0.1
Ethanone-1-(1,4-dimethyl -3-cyclohexen-1-yl)	n.d.	n.d.	0.5	n.d.	n.d.
trans-Verbenol	n.d.	n.d.	n.d.	n.d.	0.3
Borneol	n.d.	n.d.	n.d.	n.d.	tr
Terpinen-1-ol	n.d.	n.d.	0.2	n.d.	n.d.
Terpinen-4-ol	0.7	n.d.	2.4	0.4	0.9
p-Cymene-8-ol	0.4	n.d.	0.1	n.d.	n.d.
α-Terpineol	n.d.	n.d.	n.d.	0.1	tr
Fenchone	1.1	n.d.	n.d.	n.d.	n.d.
4,5-Epoxy-1-isopropyl-4-methyl-1-Cyclohexene	n.d.	n.d.	0.9	n.d.	n.d.
Dihydrocarvone	0.3	n.d.	n.d.	n.d.	0.7
Carvone	4.0	n.d.	0.2	n.d.	0.2
Thymoquinone	0.6	3.8	11.3	n.d.	tr
Thymohydroquinone	n.d.	n.d.	n.d.	n.d.	tr
β-Caryophyllene	n.d.	n.d.	0.1	n.d.	tr
Bornyl acetate	n.d.	n.d.	0.4	n.d.	0.1
2-Tridecanone	n.d.	n.d.	0.1	n.d.	n.d.

	Nickavar et al., (2003)	Moretti et al., (2004)	Singh et al., (2005)	Hamrouni et al., (2008)	Wajs et al., (2008)
2-Undecanone	n.d.	n.d.	0.1	n.d.	n.d.
Thymol	n.d.	26.8	0.1	1.8	tr
Carvacrol	1.6	n.d.	2.1	n.d.	0.3
α-Longipinene	0.3	n.d.	1.5	n.d.	0.1
Cyclosativene	n.d.	n.d.	n.d.	n.d.	tr
α-Copaene	n.d.	n.d.	n.d.	n.d.	tr
Octen-3-ol	n.d.	n.d.	n.d.	6.5	n.d.
Linalol	n.d.	n.d.	0.1	0.1	n.d.
Longicyclene	n.d.	n.d.	n.d.	n.d.	0.4
Longifolene	0.7	3.1	6.3	n.d.	tr
allo-Isolongifolene	n.d.	n.d.	n.d.	n.d.	tr
β-cis-Farnesene	n.d.	n.d.	n.d.	n.d.	tr
α-Selinene	n.d.	2.2	n.d.	n.d.	n.d.
β-Selinene	n.d.	0.4	n.d.	n.d.	0.1
7-epi-α-Selinene	n.d.	0.3	n.d.	n.d.	n.d.
β-Elemene	n.d.	5.5	n.d.	n.d.	n.d.
α-Cubebene	n.d.	n.d.	n.d.	n.d.	n.d.
β-Bisabolene	n.d.	n.d.	0.1	n.d.	tr
Epizonaren	n.d.	n.d.	0.1	n.d.	n.d.
Sesquiterpene alcohol acetate	n.d.	n.d.	0.2	n.d.	n.d.
Pimara-8(14),15-diene	n.d.	n.d.	0.2	n.d.	n.d.
γ-Cadinene	n.d.	n.d.	n.d.	n.d.	0.1
Tridecan-2-one	n.d.	n.d.	n.d.	n.d.	0.1
Farnesal D	n.d.	n.d.	n.d.	n.d.	0.2
1,8-Cineole	n.d.	n.d.	tr	1.9	n.d.
n-Nonane	1.7	n.d.	n.d.	n.d.	n.d.
3-Methyl nonane	0.3	n.d.	n.d.	n.d.	n.d.
1,3,5-Trimethyl benzene	0.5	n.d.	n.d.	n.d.	n.d.
n-Decane	0.4	n.d.	n.d.	n.d.	n.d.
1-Methyl-3-propyl benzene	0.5	n.d.	n.d.	n.d.	n.d.
1-Ethyl-2,3-dimethyl benzene	0.2	n.d.	n.d.	n.d.	n.d.
n-Tetradecane	0.2	n.d.	n.d.	n.d.	n.d.
n-Hexadecane	0.2	n.d.	n.d.	n.d.	n.d.
Estragole	1.9	n.d.	n.d.	n.d.	n.d.
Anisaldehyde	1.7	n.d.	n.d.	n.d.	n.d.
trans-Anethole	38.3	n.d.	0.6	n.d.	n.d.
Myristicin	1.4	n.d.	n.d.	n.d.	tr
Pentadecan-2-one	n.d.	n.d.	n.d.	n.d.	tr
Dill apiole	1.8	n.d.	n.d.	n.d.	n.d.
Apiole	1.0	n.d.	n.d.	n.d.	0.1

[a] n. d. = not determined. [b] tr = traces.

Wajs et al., (2008) identified forty-eight compounds, fourteen of which were volatiles that had never been isolated before. The main identified constituents were: p-cymene (60.2%), γ-terpinene (12.9%), α-thujene (7.2%), carvacrol (3.0%), α-pinene (2.0%) and β-pinene (2.1%). The major components were monoterpenes (87.7%) and their oxygenated derivatives (9.9%), while sesquiterpenes and their oxygenated derivatives constituted only a small fraction of the essential oils (0.7% and 0.3%, respectively).

NIGELLA SEEDS AND THEIR BYPRODUCTS AS FEEDS FOR ANIMAL

Because of their high nutritive value, BC seeds, and their oil and byproducts can be used as flexible ingredients to formulate balanced rations for livestock.

Different studies have shown a beneficial effect of using BC seed, oil- or meal-product supplementations in diets for ruminants on most productive and reproductive parameters (Awadalla, 1997; Gabr et al., 1998; Yousef et al., 1998; Zaki et al., 1998; El-Ayek et al., 1999; El-Ayek, 1999; El-Ekhnawy et al., 1999; Badawy et al., 2001; El-Gendy et al., 2001; El-Kady et al., 2001; Khattab et al., 2001; Kholif and Abd El-Gawad, 2001; Yousef and Zaki, 2001; Abd El-Ghani, 2003; El-Gaafarawy et al., 2003; Ibrahim et al., 2003; Mohamed et al., 2003; Salem et al., 2004; Kumari and Akbar, 2006; Abdel-Magid et al., 2007; Abo El-Nor et al., 2007; El-Saadany et al., 2008; Abo-Donia et al., 2009; Bhatt et al., 2009; Hassan et al., 2010; Khattab et al., 2011; Habeeb and El-Tarabany, 2012. Mirzaei et al., 2012; Zanouny et al., 2013a, 2013b; El-Halim et al., 2014; Cherif et al., 2018).

Several researches have been conducted on BC used as a growth promoter alone or in association with other spices with the aim of improving the performance of laying hens (Akhtar et al., 2003; Denli et al., 2004; Aydin et al., 2006; El-Bagir et al., 2006; Yalçın et al., 2009; Islam et al., 2011) and broiler chickens (Osman and El-Barody, 1999; Al-Homidan et al., 2002; Abaza et al., 2003; Guler et al., 2006; Durrani et al., 2007; Abu-Dieyeh and Abu-Darwish, 2008; Al-Beitawi and El-Ghousein, 2008;

Al-Beitawi et al., 2009; Ashayerizadeh et al., 2009; Erener et al., 2010; Nasir and Grashorn, 2010; Toghyani et al., 2010; Shewita and Taha, 2011; Hermes et al., 2011; Waheed et al., 2017. Kadhim et al., 2018).

Finally, the effects of BC have also been evaluated on the productive and reproductive performances, mortality, hematological factors, immune response, carcass traits, digestibility and caecal microbial activity of rabbits (Amber et al., 2001; Abd El- Hakim et al., 2002; El-Wafa et al., 2002; Radwan, 2002; Ismail et al., 2003; Daader et al., 2004; El-Ayek et al., 2004; El-Tohamy et al., 2007; Asgary et al., 2008; Sogut et al., 2008; Zeweil et al., 2008; Marai et al., 2009; Abdullah and Al-Kuhla, 2010; El-Bagir et al., 2010a, 2010b; El-Tohamy et al., 2010; Ali at al., 2011; Merez et al., 2011; Tousson et al., 2011; Asgary et al., 2013; Gaafar et al., 2014a, 2014b).

Conclusion

The nutritive value of black cumin is due to its carbohydrate, fatty acid, protein and several bioactive compound contents. The seeds or their byproducts can be used as feeds for farm animals, with positive effects on the compositional characteristics of eggs, milk and meat.

References

Abaza, I. M., Asar, M. A., Elshaarrawi, G. E. & Hassan, M. F. (2003). Effect of using nigella seeds, chamomile flowers, thyme flowers and harmala seeds as feed additives on performance of broiler. *Egyptian Journal of Agricultural Research*, *81*, 735-750.

Abdel-Aal, E. S. M. & Attia, R. S. (1993). Characterization of black cumin (*Nigella sativa*) seeds. 1-Chemical composition and lipids. *Alexandria Science Exchange*, *14*, 467-482.

Abd El-Ghani, M. H. (2003). Effect of cumin seed meal (*Nigella sativa*) as feed ingredient in growing lambs. *Egyptian Journal of Nutrition and Feeds*, *6*, 49-57.

Abd El-Hakim, A. S., Sedki, A. A. & Ismail, A. M. (2002). Black seed forms and its effect on rabbit performance and blood constituents. In: *Proceedings of the 3rd Scientific Conference Rabbit Production in Hot Climates*, Hurghada, Egypt, pp. 579-588.

Abdel-Magid, S. S., El-Kady, R. I., Gad, S. M. & Awadalla, I. M. (2007). Using cheep and local non-conventional protein meal (*Nigella sativa*) as least cost rations formula on performance of crossbreed calves. *International Journal of Agriculture and Biology*, *9*, 877-880.

Abdel-Wahhab, M. A. & Aly, S. E. (2005). Antioxidant property of *Nigella sativa* (black cumin) and *Syzygium aromaticum* (clove) in rats during aflatoxicosis. *Journal of Applied Toxicology: An International Journal*, *25*, 218-223.

Abdullah, N. M. & Al-Kuhla, A. A. (2010). The effect of substituting *Nigella sativa* meal as a source of protein in the rations of local rabbits on their productive performance and carcass traits. *Iraqi Journal of Veterinary Sciences*, *24*, 59-63.

Abo-Donia, F. M., Afify, A. A., Osman, A. O. & Yousef, M. M. (2009). Effect of added *Punica granatum* peel fruits and *Nigella sativa* seeds on immunology and performance of suckling buffalo calves. In: *FAO/IAEA International Symposium on Sustainable Improvement of Animal Production and Health*, Vienna, Austria, pp. 180-181.

Abo El-Nor, S. A. H., Khattab, H. M., Al-Alamy, H. A., Salem, F. A. & Abdou, M. M. (2007). Effect of some medicinal plants seeds in the rations on the productive performance of lactating buffaloes. *International Journal of Dairy Science*, *2*, 348-355.

Abu-Dieyeh, Z. H. M. & Abu-Darwish, M. S. (2008). Effect of feeding powdered black cumin seeds (*Nigella sativa* L.) on growth performance of 4-8 week-old broilers. *Journal of Animal and Veterinary Advances*, *7*, 286-290.

Ahmad, A., Husain, A., Mujeeb, M., Khan, S. A., Najmi, A. K., Siddique, N. A., Damanhouri, Z. A. & Anwar, F. (2013). A review on therapeutic potential of *Nigella sativa*: A miracle herb. *Asian Pacific Journal of Tropical Biomedicine*, *3*, 337-352.

Ahmad, M. N. & Qasem, M. A. (2017). The effect of dietary raw and roasted *Nigella sativa* L. on streptozotocin-induced changes of serum glucose and body weight in rats. *Research Journal of Pharmaceutical, Biological and Chemical Sciences*, *8*, 1298-1306.

Akhondian, J., Parsa, A. & Rakhshande, H. (2007). The effect of *Nigella sativa* L. (black cumin seed) on intractable pediatric seizures. *Medical Science Monitor*, *13*, CR555-CR559.

Akhtar, M. S., Nasir, Z. & Abid, A. R. (2003). Effect of feeding powdered *Nigella sativa* L. seeds on poultry egg production and their suitability for human consumption. *Veterinarski Arhiv*, *73*, 181-190.

Al-Beitawi, N. A. & El-Ghousein, S. S. (2008). Effect of feeding different levels of *Nigella sativa* seeds (black cumin) on performance, blood constituents and carcass characteristics of broiler chicks. *International Journal of Poultry Science*, *7*, 715-721.

Al-Beitawi, N. A., El-Ghousein, S. S. & Nofal, A. H. (2009). Replacing bacitracin methylene disalicylate by crushed *Nigella sativa* seeds in broiler rations and its effects on growth, blood constituents and immunity. *Livestock Science*, *125*, 304-307.

Alenzi, F. Q., Altamimi, M. A. A., Kujan, O., Tarakji, B., Tamimi, W., Bagader, O., Al-Shangiti, A., Talohi, A. N., Alenezy, A. K., Al-Swailmi, F., Alenizi, D., Salem, M. L. & Wyse, R. K. H. (2013). Antioxidant properties of *Nigella sativa*. *Journal of Molecular and Genetic Medicine*, *7*, 1-5.

Al-Gaby, A. M. (1998). Amino acid composition and biological effects of supplementing broad bean and corn proteins with *Nigella sativa* (black cumin) cake protein. *Nahrung*, *42*, 290-294.

Al-Homidan, A., Al-Qarawi, A. A., Al-Waily, S. A. & Adam, S. E. (2002). Response of broiler chicks to dietary *Rhazya stricta* and *Nigella sativa*. *British Poultry Science*, *43*, 291-296.

Ali, B. H. & Blunden, G. (2003). Pharmacological and toxicological properties of *Nigella sativa*. *Phytotherapy Research*, *17*, 299-305.

Ali, F. A. F., Omer, H. A. A., Abedo, A. A., Abdel-Magid, S. S. & Ibrahim, S. A. M. (2011). Using mixture of sweet basal and black cumin as feed additives with different levels of energy in growing rabbit diets. *American-Eurasian Journal of Agricultural & Environmental Sciences*, *10*, 917-927.

Aljabre, S. H., Randhawa, M. A., Akhtar, N., Alakloby, O. M., Alqurashi, A. M. & Aldossary, A. (2005). Antidermatophyte activity of ether extract of *Nigella sativa* and its active principle, thymoquinone. *Journal of Ethnopharmacology*, *101*, 116-119.

Aljabre, S. H., Alakloby, O. M. & Randhawa, M. A. (2015). Dermatological effects of *Nigella sativa*. *Journal of Dermatology & Dermatologic Surgery*, *19*, 92-98.

Al-Jassir, M. S. (1992). Chemical composition and microflora of black cumin (*Nigella sativa* L.) seeds growing in Saudi Arabia. *Food Chemistry*, *45*, 239-242.

Al-Naqeep, G., Al-Zubairi, A. S., Ismail, M., Amom, Z. H. & Esa, N. M. (2011). Antiatherogenic potential of *Nigella sativa* seeds and oil in diet-induced hypercholesterolemia in rabbits. *Evidence-Based Complementary and Alternative Medicine*, *Vol. 2011*, Article ID 213628, 8 pages.

Al-Othman, A. M., Ahmad, F., Al-Orf, S., Al-Murshed, K. S. & Ariff, Z. (2006). Effect of dietary supplementation of *Ellataria cardamun* and *Nigella sativa* on the toxicity of rancid corn oil in rats. *International Journal of Pharmacology*, 2, 60-65.

Amber, K., Abou-Zeid, A. E. & Osman, M. (2001). Influence of replacing *Nigella sativa* cake for soybean meal on growth performance and caecal microbial activity of weanling New Zealand White rabbits. *Egyptian Journal of Rabbit Science*, 11, 191-206.

Amin, B. & Hosseinzadeh, H. (2016). Black cumin (*Nigella sativa*) and its active constituent, thymoquinone: an overview on the analgesic and anti-inflammatory effects. *Planta Medica*, *82*, 8-16.

Asgary, S., Dinani, N. J., Ghanadi, A. & Helalat, A. (2008). Study the effect of *Nigella sativa* L. on hematological factors in normal and hypercholesterolemic rabbits. *Iranian Journal of Medicinal and Aromatic Plants*, *24*, 66-73.

Asgary, S., Ghannadi, A., Dashti, G., Helalat, A., Sahebkar, A. & Najafi, S. (2013). *Nigella sativa* L. improves lipid profile and prevents atherosclerosis: evidence from an experimental study on hypercholesterolemic rabbits. *Journal of Functional Foods*, *5*, 228-234.

Ashayerizadeh, O., Dastar, B., Shams Shargh, M., Ashayerizadeh, A., Rahmatnejad, E. & Hossaini, S. M. R. (2009). Use of garlic (*Allium sativum*), black cumin seeds (*Nigella sativa* L.) and wild mint (*Mentha longifolia*) in broiler chickens diets. *Journal of Animal and Veterinary Advances*, *8*, 1860-1863.

Ashraf, M., Ali, Q. & Iqbal, Z. (2006). Effect of nitrogen application rate on the content and composition of oil, essential oil and minerals in black cumin (*Nigella sativa* L.) seeds. *Journal of the Science of Food and Agriculture*, *86*, 871-876.

Atta, M. B. & Imaizumi, K. (1998). Antioxidant activity of nigella (*Nigella sativa* L.) seeds extracts. *Journal of Japan Oil Chemists' Society*, *47*, 475-480.

Atta, M. B. (2003). Some characteristics of nigella (*Nigella sativa* L.) seeds cultivated in Egypt and its lipid profile. *Food Chemistry*, *83*, 63-68.

Awadalla, I. M. (1997). The use of black cumin seed (*Nigella sativa*) cake in rations of growing sheep. *Egyptian Journal of Nutrition and Feeds*, *1*, 243-249.

Awadalla, I. M. & Gehad, A. E. (2003). Effect of supplementing growing sheep rations with black cumin seeds (*Nigella sativa*). *Journal of Agricultural Science - Mansoura University*, *28*, 185-194.

Aydin, R., Bal, M. A., Ozuğur, A. K., Toprak, H. H. C., Kamalak, A. & Karaman, M. (2006). Effects of black seed (*Nigella sativa* L.) supplementation on feed efficiency, egg yield parameters and shell quality in chickens. *Pakistan Journal of Biological Sciences*, *9*, 243-247.

Azeem, T., Zaib-Ur-Rehman, U. S., Asif, M., Arif, M. & Rahman, A. (2014). Effect of *Nigella Sativa* on poultry health and production: a review. *Science Letter*, *2*, 76-82.

Badawy, S. A., Darwish, G. M. & Rakha, G. M. (2001). Possible effects of supplementation with *Nigella sativa* cake on reproductive performance ovarian response and embryo recovery of female baladi goats. *Veterinary Medical Journal-Giza*, *49*, 507-522.

Baghdadi, H. B. & Al-Mathal, E. M. (2011). Anti-coccidial activity of *Nigella sativa* L. *Journal of Food, Agriculture and Environment*, *9*, 10-17.

Bamosa, A. O., Ali, B. A. & Al-Hawsawi, Z. A. (2002). The effect of thymoquinone on blood lipids in rats. *Indian Journal of Physiology and Pharmacology*, *46*, 195-201.

Bamosa, A. O. (2015). A review on the hypoglycemic effect of *Nigella sativa* and thymoquinone. *Saudi Journal of Medicine and Medical Sciences*, *3*, 2-7.

Beheshti, F., Khazaei, M. & Hosseini, M. (2016). Neuropharmacological effects of *Nigella sativa*. *Avicenna Journal of Phytomedicine*, *6*, 104-116.

Bhatt, N., Singh, M. & Ali, A. (2009). Effect of feeding herbal preparations on milk yield and rumen parameters in lactating crossbred cows. *International Journal of Agriculture and Biology*, *11*, 721-726.

Bourgou, S., Pichette, A., Marzouk, B. & Legault, J. (2012). Antioxidant, antiinflammatory, anticancer and antibacterial activities of extracts from *Nigella Sativa* (Black Cumin) plant parts. *Journal of Food Biochemistry*, *36*, 539-546.

Burits, M. & Bucar, F. (2000). Antioxidant activity of *Nigella sativa* essential oil. *Phytotherapy Research*, *14*, 323-328.

Chaieb, K., Kouidhi, B., Jrah, H., Mahdouani, K. & Bakhrouf, A. (2011). Antibacterial activity of Thymoquinone, an active principle of *Nigella sativa* and its potency to prevent bacterial biofilm formation. *BMC Complementary and Alternative Medicine*, *11*, 29-35.

Cheikh-Rouhou, S., Besbes, S., Hentati, B., Blecker, C., Deroanne, C. & Attia, H. (2007). *Nigella sativa* L.: Chemical composition and physicochemical characteristics of lipid fraction. *Food Chemistry*, *101*, 673-681.

Cherif, M., Valenti, B., Abidi, S., Luciano, G., Mattioli, S., Pauselli, M., Bouzarraa, I., Priolo, A. & Salem, H. B. (2018). Supplementation of *Nigella sativa* seeds to Barbarine lambs raised on low-or high-concentrate diets: effects on meat fatty acid composition and oxidative stability. *Meat Science*, *139*, 134-141.

Daader, A. H., Nasr-Alla, M. M., Azazi, I. A., Attia, S. A. M. & Seleem, T. S. T. (2004). Amelioration of heat stressed Bouscat rabbits by feeding diurnally or nocturnally diets containing *Nigella sativa* L. or fenugreek. World Rabbit Science, 12, 199.

Datta, A. K., Saha, A., Bhattacharya, A., Mandal, A., Paul, R. & Sengupta, S. (2012). Black cumin (*Nigella sativa* L.) - a review. *Journal of Plant Development Sciences*, *4*, 1-43.

Denli, M., Okan, F. & Uluocak, A. N. (2004). Effect of dietary black seed (*Nigella sativa* L.) extract supplementation on laying performance and egg quality of quail (*Coturnix coturnix japonica*). *Journal of Applied Animal Research*, *26*, 73-76.

Durrani, F. R., Chand, N., Zaka, K., Sultan, A., Khattak, F. M. & Durrani Z. (2007). Effect of different levels of feed added black seed (*Nigella sativa* L.) on the performance of broiler chicks. *Pakistan Journal of Biological Sciences*, *10*, 4164-4167.

El-Ayek, M. Y., Gabr, A. A. & Mehrez, A. Z. (1999). Influence of substituting concentrate feed mixture by *Nigella sativa* meal on animal performance and carcass traits of growing lambs. *Egyptian Journal of Nutrition and Feeds*, 2, 265-277.

El-Ayek, M. Y. (1999). Influence of substituting concentrate feed mixture by *Nigella sativa* meal on: 1-Voluntary intake, digestibility, some rumen parameters and microbial protein yield with sheep. *Egyptian Journal of Nutrition and Feeds*, 2, 279-296.

El-Ayek, M. Y., El-Harairy, M. A. & Mousa, M. O. (2004). Influence of substituting concentrate feed mixture by *Nigella sativa* meal on the performance of growing rabbits. *World Rabbit Science*, *12*, 212 (Abst.).

El-Bagir, N. M., Hama, A. Y., Hamed, R. M., Abd El Rahim, A. G. & Beynen, A. C. (2006). Lipid composition of egg yolk and serum in laying hens fed diets containing black cumin (*Nigella sativa*). *International Journal of Poultry Science*, *5*, 574-578.

El-Bagir, N. M., Farah, I. T. O., Alhaidary, A., Mohamed, H. E. & Beynen, A. C. (2010a). Clinical laboratory serum values in rabbits fed diets containing black cumin seeds. *Journal of Animal and Veterinary Advances*, *9*, 2532-2536.

El-Bagir, N. M., Farah, I. T. O., Elhag, S. M. B., Alhaidary, A., Mohamed, H. E. & Beynen, A. C. (2010b). Immune response and *Pasteurella* resistance in rabbits fed diets containing various amounts of black cumin seeds. *American Journal of Animal and Veterinary Sciences*, *5*, 163-167.

El-Ekhnawy, K. E., Otteifa, A. M., Ezzo, O. H. & Hegazy, M. A. (1999). Post weaning reproductive activity of Barki ewes lambing in spring fed *Nigella sativa* oil seed meal. *Assiut Veterinary Medical Journal*, *40*, 292-309.

El-Gaafarawy, A. M., Zaki, A. A., El-Sedly, E. R. & El-Ekhnawy, I. K. (2003). Effect of feeding *Nigella sativa* cake on digestibility, nutritive value and reproductive performance of Frisian cows and immuno activity of their offspring. *Egyptian Journal of Nutrition and Feeds*, *6*, 307-313.

El-Gendy, K. M., Zaki, A. A., Abou Ammo, F. F. & El-Gamal, M. F. A. (2001). *Nigella sativa* meal as a protein supplement in ruminant rations. *Egyptian Journal of Nutrition and Feeds*, *4*, 1-7.

El-Gendy, S., Hessien, M., Abdel Salam, I., Morad, M., EL-Magraby, K., Ibrahim, H. A., Kalifa, M. H. & El-Aaser, A. A. (2007). Evaluation of the possible antioxidant effects of soybean and *Nigella sativa* during experimental hepatocarcinogenesis by nitrosamine precursors. *Turkish Journal of Biochemistry*, *32*, 5-11.

El-Halim, M. I. A., El-Bagir N. M. & Sabahelkhier, M. K. (2014). Hematological values in sheep fed a diet containing black cumin (*Nigella sativa*) seed oil. *International Journal of Biochemistry Research & Review*, *4*, 128-140.

El-Kady, R. I., Kandiel, A. M. & Etman, A. H. (2001). Effect of substituting concentrate-protein by *Nigella sativa* meal on growing calves performance. *Journal of Agricultural Science - Mansoura University*, *12*, 7645-7655.

El-Komey, A. G. (1996). Effect of black seed (*Nigella sativa*) during pregnancy and lactation on mammary gland development in rat. *Alexandria Journal of Agricultural Sciences*, *41*, 63-74.

El-Saadany, S. A., Habeeb, A. A. M., El-Gohary, E. S., El-Deeb, M. M. & Aiad, K. M. (2008). Effect of supplementation of oregano or *Nigella sativa* seeds to diets of lactating Zaraibi goats on milk yield and some physiological functions during summer season. *Egyptian Journal of Animal Production*, *45*, 469-487.

El-Shenawy, N. S., Soliman, M. F. N. & Reyad, S. I. (2008). The effect of antioxidant properties of aqueous garlic extract and *Nigella sativa* as anti-schistosomiasis agents in mice. *Revista do Instituto de Medicina Tropical de São Paulo*, *50*, 29-36.

El-Tahir, K. E. D. H. & Bakeet, D. M. (2006). The black seed *Nigella sativa* Linnaeus - A mine for multi cures: a plea for urgent clinical evaluation of its volatile oil. *Journal of Taibah University Medical Sciences*, *1*, 1-19.

El-Tohamy, M. M. & El-Kady, R. I. (2007). Partial replacement of soybean meal with some medicinal plant seed meals and their effect on the performance of rabbits. *International Journal of Agriculture and Biology*, *9*, 215-219.

El-Tohamy, M. M., El-Nattat, W. S. & El-Kady R. I. (2010). The beneficial effects of *Nigella sativa*, *Raphanus sativus* and *Eruca sativa* seed cakes to improve male rabbit fertility, immunity and production. *Journal of American Science*, *6*, 1247-1255.

El-Wafa, S. A., Sedki, A. A. & Ismail, A. M. (2002). Response of growing rabbits to diets containing black seed, garlic or onion as natural feed additives. *Egyptian Journal of Rabbit Science*, *12*, 69-83.

Erener, G., Altop, A., Ocak, N., Aksoy, H. M., Cankaya, S. & Ozturk E. (2010). Influence of black cumin seeds (*Nigella sativa* L.) and seed extract on broilers performance and total coliform bacteria count. *Asian Journal of Animal and Veterinary Advances*, *5*, 128-135.

Erkan, N., Ayranci, G. & Ayranci, E. (2008). Antioxidant activities of rosemary (*Rosmarinus officinalis* L.) extract, blackseed (*Nigella sativa* L.) essential oil, carnosic acid, rosmarinic acid and sesamol. *Food Chemistry*, *110*, 76-82.

Forouzanfar, F., Bazzaz, B. S. F. & Hosseinzadeh, H. (2014). Black cumin (*Nigella sativa*) and its constituent (thymoquinone): a review on antimicrobial effects. *Iranian Journal of Basic Medical Sciences*, *17*, 929-938.

Gaafar, H. M. A., Ragab, A. A. & El-Reidy, K. F. A. (2014a). Effect of diet supplemented with pumpkin (*Cucurbita moschata*) and black seed (*Nigella sativa*) oils on performance of rabbits: 1 - Growth performance, blood haematology and carcass traits of growing rabbits. *Report and Opinion*, *6*, 52-59.

Gaafar, H. M. A., Ragab, A. A. & El-Reidy K. F. A. (2014b). Effect of diet supplemented with pumpkin (*Cucurbita moschata*) and black seed (*Nigella sativa*) oils on performance of rabbits: 2 - Productive and reproductive performance of does and their offspring. *Report and Opinion*, *6*, 60-68.

Gabr, A. A., El-Ayouty, S. A., Zaki, A. A., Abou Ammo, F. F. & El-Gohary, E. S. I. (1998). Productive performance of lambs fed with diets containing *Nigella sativa* meal. *Egyptian Journal of Nutrition and Feeds*, *1*, 97-107.

Gali-Muhtasib, H., El-Najjar, N. & Schneider-Stock, R. (2006). The medicinal potential of black seed (*Nigella sativa*) and its components. *Advances in Phytomedicine*, *2*, 133-153.

Gholamnezhad, Z., Keyhanmanesh, R. & Boskabady, M. H. (2015). Anti-inflammatory, antioxidant, and immunomodulatory aspects of *Nigella sativa* for its preventive and bronchodilatory effects on obstructive respiratory diseases: A review of basic and clinical evidence. *Journal of Functional Foods*, *17*, 910-927.

Gholamnezhad, Z., Havakhah, S. & Boskabady, M. H. (2016). Preclinical and clinical effects of *Nigella sativa* and its constituent, thymoquinone: A review. *Journal of Ethnopharmacology*, *190*, 372-386.

Gilani, A. U. H., Jabeen, Q. & Khan, M. A. U. (2004). A review of medicinal uses and pharmacological activities of *Nigella sativa*. *Pakistan Journal of Biological Sciences*, *7*, 441-451.

Guler, T., Dalkılıç, B., Ertas, O. N. & Çiftçi, M. (2006). The effect of dietary black cumin seeds (*Nigella sativa* L.) on the performance of broilers. *Asian-Australasian Journal of Animal Sciences*, *19*, 425-430.

Habeeb, A. A. M. & El-Tarabany, A. A. (2012). Effect of *Nigella sativa* or curcumin on daily body weight gain, feed intake and some physiological functions in growing Zaraibi goats during hot summer season. *Arab Journal of Nuclear Sciences and Application*, *45*, 238-249.

Hamrouni-Sellami, I., Elyes Kchouk, M. & Marzouk, B. (2008). Lipid and aroma composition of black cumin (*Nigella sativa* L.) seeds from Tunisia. *Journal of Food Biochemistry*, *32*, 335-352.

Harzallah, H. J., Grayaa, R., Kharoubi, W., Maaloul, A., Hammami, M. & Mahjoub, T. (2012). Thymoquinone, the *Nigella sativa* bioactive compound, prevents circulatory oxidative stress caused by 1,2-dimethylhydrazine in erythrocyte during colon postinitiation carcinogenesis. *Oxidative Medicine and Cellular Longevity*, *Vol. 2012*, Article ID 854065, 6 pages.

Hassan, S. A., Hassan, K. M. & Al-Rubeii, A. (2010). Carcass characteristics of Karadi lambs as affect by different levels of dietary supplement of rumen degradable nitrogen fed with *Nigella sativa*. *African Journal of Biotechnology*, *9*, 4295-4299.

Hayatdavoudi, P., Rad, A. K., Rajaei, Z. & Hadjzadeh, M. A. R. (2016). Renal injury, nephrolithiasis and *Nigella sativa*: A mini review. *Avicenna Journal of Phytomedicine*, *6*, 1-8.

Hermes, I. H., Attia, F. M., Ibrahim, K. A. & El-Nesr S. S. (2011). Physiological responses of broiler chickens to dietary different forms and levels of *Nigella sativa* L., during Egyptian summer season. *Journal of Agriculture and Veterinary Sciences*, *4*, 17-33.

Heshmati, J. & Namazi, N. (2015). Effects of black seed (*Nigella sativa*) on metabolic parameters in diabetes mellitus: A systematic review. *Complementary Therapies in Medicine*, *23*, 275-282.

Hosseinzadeh, H., Parvardeh, S., Asl, M. N., Sadeghnia, H. R. & Ziaee, T. (2007). Effect of thymoquinone and *Nigella sativa* seeds oil on lipid peroxidation level during global cerebral ischemia-reperfusion injury in rat hippocampus. *Phytomedicine*, *14*, 621-627.

Hussain, D. A. & Hussain, M. M. (2016). *Nigella sativa* (black seed) is an effective herbal remedy for every disease except death-a Prophetic statement which modern scientists confirm unanimously: a review. *Advancement in Medicinal Plant Research*, *4*, 27-57.

Ibrahim, G. F., Shahin, G. E. & El-Ekhnawy, K. I. (2003). Effect of adding *Nigella sativa* cake to crossbred Friesian calves feeds on productive and physiological performance. *Egyptian Journal of Applied Sciences*, *18*, 1-16.

Ijaz, H., Tulain, U. R., Qureshi, J., Danish, Z., Musayab, S., Akhtar, M. F., Saleem, A., Khan, K. A. U. R., Zaman, M., Waheed, I., Khan, I. & Abdel-Daim, M. (2017). *Nigella sativa* (Prophetic Medicine): A Review. *Pakistan Journal of Pharmaceutical Sciences*, *30*, 229-2342.

Ismail, A. M., Sedki, A. A. & Abdallah, A. G. (2003). Influence of black seed, garlic and onion supplementation on reproductive performance in rabbits. *Egyptian Journal of Agricultural Research*, *81*, 1193-1207.

Ismail, M. Y. M. & Yaheya, M. (2009). Therapeutic role of prophetic medicine Habbat El Baraka (*Nigella sativa* L.) - A review. *World Applied Sciences Journal*, 7, 1203-1208.

Ismail, M., Al-Naqeep, G. & Chan, K. W. (2010). *Nigella sativa* thymoquinone-rich fraction greatly improves plasma antioxidant capacity and expression of antioxidant genes in hypercholesterolemic rats. *Free Radical Biology and Medicine*, *48*, 664-672.

Islam, M. T., Selim, A. S. M., Sayed, M. A., Khatun, M. A., Siddiqui, M. N., Alam, M. S. & Hossain, M. A. (2011). *Nigella sativa* L. supplemented diet decreases egg cholesterol content and suppresses harmful intestinal bacteria in laying hens. *Journal of Animal and Feed Sciences*, *20*, 587-598.

Kadhim, L. I., Al-Zubaidi, M. T. S. & Al Saegh, H. A. (2018). Influence of dietary supplementation of *Nigella sativa* on experimental coccidiosis in broiler chickens. *Journal of Entomology and Zoology Studies*, *6*, 652-656.

Kanter, M., Meral, I., Dede, S., Cemek, M., Ozbek, H., Uygan, I. & Gunduz, H. (2003). Effects of *Nigella sativa* L. and *Urtica dioica* L. on lipid peroxidation, antioxidant enzyme systems and some liver enzymes in CCl_4-treated rats. *Journal of Veterinary Medicine Series A*, *50*, 264-268.

Kanter, M., Coskun, O. & Budancamanak, M. (2005). Hepatoprotective effects of *Nigella sativa* L and *Urtica dioica* L on lipid peroxidation, antioxidant enzyme systems and liver enzymes in carbon tetrachloride-treated rats. *World Journal of Gastroenterology*, *11*, 6684-6688.

Kanter, M. (2009). Effects of *Nigella sativa* seed extract on ameliorating lung tissue damage in rats after experimental pulmonary aspirations. *Acta Histochemica*, *111*, 393-403.

Khaled, A. A. S. (2009). Gastroprotective effects of *Nigella sativa* oil on the formation of stress gastritis in hypothyroidal rats. International *Journal of Physiology, Pathophysiology and Pharmacology*, *1*, 143-149.

Khan, S. A., Khan, A. M., Karim, S., Kamal, M. A., Damanhouri, G. A. & Mirza, Z. (2016). Panacea seed "Nigella": A review focusing on regenerative effects for gastric ailments. *Saudi Journal of Biological Sciences*, *23*, 542-553.

Khattab, H. M., El-Elhamy, H. A., Abo El-Nor, S. A., Salem, A. F. & Abdo, M. M. (2001). Influence effect of some medicinal plants on the performance of lactating buffaloes. *Egyptian Journal of Nutrition and Feeds*, *4*, 550-556.

Khattab, H. M., El-Basiony, A. Z., Hamdy, S. M. & Marwan, A. A. (2011). Immune response and productive performance of dairy buffaloes and their offspring supplemented with black seed oil. *Iranian Journal of Applied Animal Science*, *1*, 227-234.

Kholif, A. M. & Abd EI-Gawad, M. A. M. (2001). Medicinal plant seed supplementation of lactating goat diets and its effects on milk and cheese quality and quantity. *Egyptian Journal of Dairy Science*, *29*, 139-150.

Kooti, W., Hasanzadeh-Noohi, Z., Sharafi-Ahvazi, N., Asadi-Samani, M. & Ashtary-Larky, D. (2016). Phytochemistry, pharmacology, and therapeutic uses of black seed (*Nigella sativa*). *Chinese Journal of Natural Medicines*, *14*, 732-745.

Kumari, R. & Akbar, M. A. (2006). Clinical efficacy of some herbal drugs during indigestion in Buffaloes. *Buffalo Bulletin*, *25*, 3-6.

Longato, E., Meineri, G. & Peiretti, P. G. (2015). Nutritional and zootechnical aspects of *Nigella sativa*: A review. *Journal of Animal and Plant Sciences*, *25*, 921-934.

Mahdavi, R., Heshmati, J. & Namazi, N. (2015). Effects of black seeds (*Nigella sativa*) on male infertility: A systematic review. *Journal of Herbal Medicine*, *5*, 133-139.

Majdalawieh, A. F. & Fayyad, M. W. (2015). Immunomodulatory and anti-inflammatory action of *Nigella sativa* and thymoquinone: A comprehensive review. *International Immunopharmacology*, *28*, 295-304.

Mansour, M. A., Ginawi, O. T., El-Hadiyah, T., El-Khatib, A. S., Al-Shabanah, O. A. & Al-Sawaf, H. A. (2001). Effects of volatile oil constituents of *Nigella sativa* on carbon tetrachloride-induced hepatotoxicity in mice: evidence for antioxidant effects of thymoquinone. *Research Communications in Molecular Pathology and Pharmacology*, *110*, 239-251.

Maqbool, A., Sikandar Hayat, C. & Tanveer, A. (2004). Comparative efficacy of various indigenous and allopathic drugs against fasciolosis in buffaloes. *Veterinarski Arhiv*, *74*, 107-114.

Marai, I. F. M., Abdel-Monem, U. M. & Soliman, M. M. (2009). Effect of feeding system and dietary *Nigella sativa* seed level on performance of New Zealand White rabbit does during the mild and hot seasons in Egypt. *Egyptian Journal of Rabbit Science*, *19*, 37-50.

Mariod, A. A., Edris, Y. A., Cheng, S. F. & Abdelwahab, S. I. (2012). Effect of germination periods and conditions on chemical composition, fatty acids and amino acids of two black cumin seeds. *Acta Scientiarum Polonorum Technologia Alimentaria*, *11*, 401-410.

Mazaheri, Y., Torbati, M., Azadmard-Damirchi, S. & Savage, G. P. (2019). A comprehensive review of the physicochemical, quality and nutritional properties of *Nigella sativa* oil. *Food Reviews International*, *35*, 342-362.

Merez, A. Z., El-Harairy, M. A. & Salama, M. M. M. (2011). Effect of using black seed on growth performance and economical efficiency of rabbits. *Journal of Animal and Poultry Production - Mansoura University*, *2*, 13-21.

Mirzaei, F. (2012). Effect of herbal feed additives on performance parameters of ruminants and especially on dairy goat: a review. *International Journal of Agricultural Sciences and Veterinary Medicine*, *6*, 307-331.

Mirzaei, F., Prasad, S. & Savar Sofla, S. (2012). Influence of medicinal plants mixture on productive performance cross bred dairy goats. *Current Research in Dairy Sciences*, *4*, 6-16.

Mohamed, A. H., El-Saidy, B. E. & Seidi, I. A. (2003). Influence of some medicinal plants supplementation: 1- On d igestibility, nutritive values, rumen fermentation parameters in sheep. *Egyptian Journal of Nutrition and Feeds*, *6*, 139-150.

Mohtashami, A. & Entezari, M. H. (2016). Effects of *Nigella sativa* supplementation on blood parameters and anthropometric indices in adults: A systematic review on clinical trials. *Journal of Research in Medical Sciences*, *21*, 3.

Mollazadeh, H. & Hosseinzadeh, H. (2014). The protective effect of *Nigella sativa* against liver injury: a review. *Iranian Journal of Basic Medical Sciences*, *17*, 958-966.

Moretti, A., D'Antuono, L. F. & Elementi, S. (2004). Essential oils of *Nigella sativa* L. and *Nigella damascena* L. seeds. *Journal of Essential Oil Research*, *16*, 182-183.

Morsi, N. M. (2000). Antimicrobial effect of crude extracts of *Nigella sativa* on multiple antibiotics-resistant bacteria. *Acta Microbiologica Polonica*, *49*, 63-74.

Nagi, M. N., Alam, K., Badary, O. A., Al-Shabanah, O. A., Al-Sawaf, H. A. & Al-Bekairi, A. M. (1999). Thymoquinone protects against carbon tetrachloride hepatotoxicity in mice via an antioxidant mechanism. *Biochemistry and Molecular Biology International*, *47*, 153-159.

Nasir, Z. & Grashorn, M. A. (2010). Effects of *Echinacea purpurea* and *Nigella sativa* supplementation on broiler performance, carcass and meat quality. *Journal of Animal and Feed Sciences*, *19*, 94-104.

Naz, H. (2011). *Nigella sativa*: the miraculous herb. *Pakistan Journal of Biochemistry and Molecular Biology*, *44*, 44-48.

Nickavar, B., Mojab, F., Javidnia, K. & Roodgar Amoli, M. A. (2003). Chemical composition of the fixed and volatile oils of *Nigella sativa* L. from Iran. *Zeitschrift für Naturforschung*, *58*, 629-631.

Osman, A. M. A. & El-Barody, M. A. A. (1999). Growth performance and immune response of broiler chicks as affected by diet density and *Nigella sativa* seed supplementation. *Egyptian Poultry Science Journal*, *19*, 619-634.

Özcan, M. (2004). Mineral contents of some plants used as condiments in Turkey. *Food Chemistry*, *84*, 437-440.

Paarakh, P. M. (2010). *Nigella sativa* Linn. – A comprehensive review. *Indian Journal of Natural Products and Resources*, *1*, 409-429.

Perveen, T., Abdullah, A., Haider, S., Sonia, B., Munawar, A. S. & Haleem, D. J. (2008). Long-term administration of *Nigella sativa* effects nociception and improves learning and memory in rats. *Pakistan Journal of Biochemistry and Molecular Biology*, *41*, 141-143.

Petrujkić, B. T., Beier, R. C., He, H., Genovese, K. J., Swaggerty, C. L., Hume, M. E., Crippen, T. L., Harvey, R. B., Anderson, R. C. & Nisbet, D. J. (2018). *Nigella sativa* L. as an alternative antibiotic feed supplement and effect on growth performance in weanling pigs. *Journal of the Science of Food and Agriculture*, *98*, 3175-3181.

Pichette, A., Marzouk, B. & Legault, J. (2012). Antioxidant, anti-inflammatory, anticancer and antibacterial activities of extracts from *Nigella sativa* (black cumin) plant parts. *Journal of Food Biochemistry*, *36*, 539-546.

Pourghassem-Gargari, B., Ebrahimzadeh-Attary, V., Rafraf, M. & Gorbani, A. (2009). Effect of dietary supplementation with *Nigella sativa* L. on serum lipid profile, lipid peroxidation and antioxidant defense system in hyperlipidemic rabbits. *Journal of Medicinal Plants Research*, *3*, 815-821.

Radwan, M. S. M. (2002). Effects of replacing soybean meal by nigella seed meal for growing rabbits on digestibility coefficients, growth performance, carcass traits and economic efficiency under hot climatic conditions. *Egyptian Journal of Rabbit Science*, *12*, 13-25.

Rahman, M. A. M. A. & Nada, A. A. (2006). Effect of black seed oil on rabbits infected with some intestinal *Eimeria* species. *Veterinary Medical Journal-Giza*, *54*, 331-341.

Ramadan, M. F. & Mörsel, J. T. (2002). Characterization of phospholipid composition of black cumin (*Nigella sativa* L.) seed oil. *Nahrung*, *46*, 240-244.

Ramadan, M. F. & Mörsel, J. T. (2003). Analysis of glycolipids from black cumin (*Nigella sativa* L.), coriander (*Coriandrum sativum* L.) and niger (*Guizotia abyssinica* Cass.) oilseeds. *Food Chemistry*, *80*, 197-204.

Ramadan, M. F. (2007). Nutritional value, functional properties and nutraceutical applications of black cumin (*Nigella sativa* L.): an overview. *International Journal of Food Science & Technology*, *42*, 1208-1218.

Randhawa, M. A. & Al-Ghamdi, M. S. (2002). A review of the pharmacotherapeutic effects of *Nigella sativa*. *Pakistan Journal of Medical Research*, *41*, 77-83.

Randhawa, M. A. (2008). Black seed, *Nigella sativa*, deserves more attention. *Journal of Ayub Medical College Abbottabad*, *20*, 1-2.

Randhawa, M. A. & Al-Ghamdi, M. S. (2011). Anticancer activity of *Nigella sativa* (black seed) - a review. *The American Journal of Chinese Medicine*, *39*, 1075-1091.

Razavi, B. M. & Hosseinzadeh, H. (2014). A review of the effects of *Nigella sativa* L. and its constituent, thymoquinone, in metabolic syndrome. *Journal of Endocrinological Investigation*, *37*, 1031-1040.

Sahebkar, A., Beccuti, G., Simental-Mendia, L. E., Nobili, V. & Bo, S. (2016a). *Nigella sativa* (black seed) effects on plasma lipid concentrations in humans: A systematic review and meta-analysis of randomized placebo-controlled trials. *Pharmacological Research, 106*, 37-50.

Sahebkar, A., Soranna, D., Liu, X., Thomopoulos, C., Simental-Mendia, L. E., Derosa, G., Maffioli, P. & Parati, G. (2016b). A systematic review and meta-analysis of randomized controlled trials investigating the effects of supplementation with *Nigella sativa* (black seed) on blood pressure. *Journal of Hypertension*, *34*, 2127-2135.

Salem, M. A. (2001). Effect of some heat treatment on nigella seeds characteristics. 1-Some physical and chemical properties of nigella seed oil. *Journal of Agricultural Research - Tanta University*, *27*, 471-486.

Salem, F. A., Abu El-Ella, A. A., Marghany, M. & El-Amary, H. (2004). Effect of medical herbs and plants as feed additives on sheep performance. *Journal of Agricultural Science - Mansoura University*, *29*, 2303.

Salem, M. L. (2005). Immunomodulatory and therapeutic proprieties of the *Nigella sativa*. *International Immunopharmacology*, *5*, 1749-1770.

Schwarz, S., Kehrenberg, C. & Walsh, T. R. (2001). Use of antimicrobial agents in veterinary medicine and food animal production. *International Journal of Antimicrobial Agents*, *17*, 431-437.

Sharma, N. K., Ahirwar, D., Jhade, D. & Gupta, S. (2009). Medicinal and phamacological potential of *Nigella sativa*: a review. *Ethnobotanical Leaflets*, *13*, 946-955.

Shewita, R. S. & Taha, A. E. (2011). Effect of dietary supplementation of different levels of black seed (*Nigella Sativa* L.) on growth performance, immunological, hematological and carcass parameters of broiler chicks. *World Academy of Science, Engineering and Technology*, *53*, 1071-1077.

Shokri, H. (2016). A review on the inhibitory potential of *Nigella sativa* against pathogenic and toxigenic fungi. *Avicenna Journal of Phytomedicine*, *6*, 21-33.

Shrivastava, R. M., Agrawal, R. C. & Parveen, Z. E. B. A. (2011). A review on therapeutic applications of *Nigella sativa. Journal of Chemistry and Chemical Sciences*, *1*, 241-248.

Singh, G., Marimuthu, P., De Heluani, C. S. & Catalan, C. (2005). Chemical constituents and antimicrobial and antioxidant potentials of essential oil and acetone extract of *Nigella sativa* seeds. *Journal of the Science of Food and Agriculture*, *85*, 2297-2306.

Sogut, B., Celik, I. & Tuluce, Y. (2008). The effects of diet supplemented with the black Cumin (*Nigella sativa* L.) upon immune potential and antioxidant marker enzymes and lipid peroxidation in broiler chicks. *Journal of Animal and Veterinary Advances*, *7*, 1196-1199.

Takruri, H. R. H. & Dameh, M. A. F. (1998). Study of the nutritional value of black cumin seeds (*Nigella sativa* L). *Journal of the Science of Food and Agriculture*, *76*, 404-410.

Tavakkoli, A., Mahdian, V., Razavi, B. M. & Hosseinzadeh, H. (2017). Review on clinical trials of black seed (*Nigella sativa*) and its active constituent, thymoquinone. *Journal of Pharmacopuncture*, *20*, 179-193.

Tembhurne, S. V., Feroz, S., More, B. H. & Sakarkar, D. M. (2014). A review on therapeutic potential of *Nigella sativa* (kalonji) seeds. *Journal of Medicinal Plants Research*, *8*, 167-177.

Toghyani, M., Toghyani, M., Gheisari, A., Ghalamkari, G. & Mohammadrezaei, M. (2010). Growth performance, serum biochemistry and blood hematology of broiler chicks fed different levels of black seed (*Nigella sativa*) and peppermint (*Mentha piperita*). *Livestock Science*, *129*, 173-178.

Tousson, E., El-Moghazy, M. & El-Atrsh, E. (2011). The possible effect of diets containing *Nigella sativa* and *Thymus vulgaris* on blood parameters and some organs structure in rabbit. *Toxicology and Industrial Health*, *27*, 107-116.

Umar, S., Zargan, J., Umar, K., Ahmad, S., Katiyar, C. K. & Khan, H. A. (2012). Modulation of the oxidative stress and inflammatory cytokine response by thymoquinone in the collagen induced arthritis in Wistar rats. *Chemico-Biological Interactions*, *197*, 40-46.

Üstun, G., Kent, L., Ḉekin, N. & Clvelekoglu, H. (1990). Investigation of the technological properties of *Nigella sativa* (Black cumin) seed oil. *Journal of the American Oil Chemists' Society*, *67*, 958-960.

Waheed, S., Hasnain, A., Ahmad, A., Tarar, O. M., Yaqeen, Z. & Ali, T. M. (2017). Effect of spices and sweet violet extracts to replace antibiotics and antioxidants in feed on broiler performance, hematology, lipid profile and immunity. *Journal of Animal and Plant Sciences*, *27*, 714-724.

Wajs, A., Bonikowski, R. & Kalemba, D. (2008). Composition of essential oil from seeds of *Nigella sativa* L. cultivated in Poland. *Flavour and Fragrance Journal*, *23*, 126-132.

Yalçın, S., Yalçın, S., Erol, H., Buğdaycı, K. E., Özsoy, B. & Çakır, S. (2009). Effects of dietary black cumin seeds (*Nigella sativa* L.) on performance, egg traits, egg cholesterol content and egg yolk fatty acid composition in laying hens. *Journal of the Science of Food and Agriculture*, *89*, 1737-1742.

Yousef, M. M., Abdiene, A. M., Khattab, R. M. & Darwish, S. A. (1998). Effect of feeding *Nigella sativa* cake on productive and reproductive performance of buffalos. *Egyptian Journal of Nutrition and Feeds*, *1*, 73-85.

Yousef, M. M. & Zaki, A. A. (2001). Effect of barley radical feeding on body weight gain and some physiological parameters of growing Friesian crossbred calves. *Egyptian Journal of Nutrition and Feeds*, *4*, 465-472.

Zaki, A. A., El-Gamal, M. F. A. & El-Gendy, K. M. (1998). Effect of feedings *Nigella sativa* meal on performance of buffaloes. *Zagazig Veterinary Journal*, *26*, 85-97.

Zanouny, A. I., Abd-Elmoty, A. K. I., Sallam, M. T., El-Barody, M. A. A. & Abd El-hakeam, A. A. (2013a). Effect of *Nigella sativa* seed supplementation on nutritive values and growth performance of Ossimi sheep. *Egyptian Journal of Sheep and Goats Sciences*, *8*, 57-63.

Zanouny, A. I., Abd-Elmoty A. K. I., El-Barody, M. A. A., Sallam, M. T. & Abd El-Hakeam, A. A. (2013b). Effect of supplementation with *Nigella sativa* seeds on some blood metabolites and reproductive performance of Ossimi male lambs. *Egyptian Journal of Sheep and Goats Sciences*, *8*, 47-56.

Zeweil, H. S., Ahmed, M. H., El-Adawy, M. M. & Zaki B. (2008). Evaluation of substituting nigella seed meal as a source of protein for soybean meal in diets of New Zealand white rabbits. In: *Proceedings of the 9th World Rabbit Congress*, Verona, Italy, pp. 863-886.

In: *Nigella sativa*
Editor: Sanjin Berghuis
ISBN: 978-1-53617-538-7

Chapter 2

ANTIOXIDANT, ANTICANCER, ANTIMICROBIAL EFFECTS OF *NIGELLA SATIVA* SEEDS AND ITS PHARMACOLOGICAL USES

Heba I. Mohamed[1,*], Mohamed Moustafa-Farag[2,3], Hossam S. El-Beltagi[4,5], Naglaa A. Ashry[6] and Eman M. Fawzi[1], Hanaa H. El-Shazly[1], Abdelfattah Badr[7], Maged M. Abou El-Enain[1]

[1]Biological and Geological Sciences Department,
Faculty of Education, Ain Shams University, Cairo, Egypt
[2] Institute of Agricultural Resources and Environment, Guangdong Academy of Agricultural Sciences, Guangzhou, Guangdong, China
[3] Horticulture Research Institute, Agriculture Research Center, Vegetable Seed Production and Technology, Giza, Egypt
[4]Biochemistry Department, Faculty of Agriculture,
Cairo University, Giza, Cairo, Egypt

[*] Corresponding Author's E-mail: hebaibrahim79@gmail.com.

[5]Agricultural Biotechnology Department,
College of Agriculture and Food Sciences,
King Faisal University, Al-Ahsa, Saudi Arabia
[6]Cell Research Department, Field Crops Research Inst.,
Agricultural Research Centre, Giza, Egypt
[7]Botany and Microbiology Department, Faculty of Science, Helwan University, Ain Helwan, Cairo, Egypt

ABSTRACT

Black cumin (*Nigella sativa* L.) belongs to Ranunculaceae family. As an aromatic plant, *N. sativa* is widely grown in different parts of the world and the seeds of black cumin have been used to promote health for Middle East and Southeast Asia countries. *N. sativa* seeds yield esters of fatty acids, free sterols and steryl esters. The seeds also contain lipase, phytosterols and sitosterol. During the past few decades, many phytochemical and pharmacological studies have been conducted on *N. sativa* seeds because of its marked biological activities, antioxidant, anti-inflammatory and antiulcer activity. The main compounds that have been isolated from *N. sativa* seed include proteins, carbohydrates, fixed oils, essential oil, crude fiber, alkaloids, minerals, vitamins, ash, and moisture. The other components are tannins, resin, saponin, carotene, glucosides and sterols. One of the major components of the essential oil is thymoquinone (TQ). Based on the use of *N. sativa* in traditional medicine as a natural treatment for some diseases, researchers have investigated its protective effects against asthma, hypertension, diabetes, and inflammation. In addition, TQ is known to have antioxidant, antifungal, antibacterial, anticancer and neuroprotective activities. It has been found that TQ has beneficial protective effects against renal diseases through anti-inflammatory, antioxidant, and antiapoptotic activities. Here, we are giving a summary of antioxidant, anticancer, antimicrobial effects of *N. sativa* on regard of pharmacological uses.

Keywords: antifungal, antibacterial, anticancer, diabetes, inflammation thymoquinone

1. INTRODUCTION

Nigella sativa L. (family *Ranunculaceae*) is commonly known as black cumin or black seed. The seed or its oil is used as a carminative, diuretic, lactagogue and vermifuge (Akgul, 1989; Ali and Blunden, 2003). The dried seeds from black cumin are also used for sprinkling on bread or flavouring foods, especially bakery products and cheese (Ustum et al., 1990; Takruri and Daneh, 1998). *N. sativa* seeds contain 36-38% fixed oils, proteins, alkaloids, saponin and 0.4-2.5% essential oil (Ali and Blunden, 2003). It has also been highlighted that the active substances of *N. sativa* have antibacterial, antifungal, antidiabetic, immunomodulator, anti-inflammatory, analgesic, antiviral, antioxidant, anticonvulsant, antihypertensive, anticancer and antihyperlipidemic effects (Hanafy and Hatem, 1991; De et al., 1999; Farrag et al., 2000; Burits and Bucar, 2000; Sagdic et al., 2002; Sagdic, 2003; Khan et al., 2003; Entok et al., 2014; Leong et al., 2013; Singh et al., 2014). Due to these effects, *N. sativa's* seed and oil have been used globally in the treatment of many diseases such as asthma, diarrhoea, dysentery, dyspepsia, fever, icterus, apoplexy, hemorrhoids and cardiovascular, digestive, immune-system, liver, respiratory and kidney diseases (Forouzanfar et al., 2014).

N. sativa has been widely used from the past to the present for various purposes, including as a painkiller and for anthelmintic as an appetiser and for carminative, sudorific, digestive, diuretic, emmenagogue, guaiacol, antifebrile, galactagouge and cathartic uses. *N. sativa* is reported to decrease asthenia and depression and to increase body resistance (Razavi and Hosseinzadeh, 2014). In this chapter we highlighted these above impacts of *N. sativa* and suggest directions for future research in this pharmacological regard.

2. *NIGELLA SATIVA* AND ITS CHEMICAL COMPOSITION

Black cumin, which is usually used as a spice, is grown as twelve different types; the most widely used type in agriculture and trade is *N.*

sativa, on which many studies have been conducted worldwide (Yakup, 2007). It is a rather pilous annual herbaceous plant with a height of approximately 20–30 cm. Its flower is five-leaved and is light or dark blue (Figure 1). The part that is used for nutrition is the seed, which consists of many white, trigonal and bitter grains with a special aroma inside the capsule (Figure 2) (Salem, 2005).

Figure 1. Flowers of *Nigella sativa.*

Figure 2. Fruit of *Nigella sativa.*

Depending on the region, the *N. sativa* seed contains volatile (0.40%–0.45%) and non-volatile (32%–40%) oils, protein (16.00%–20.85%), carbohydrates (31.0%–33.9%), fibre (5.50–7.94%), alkaloids, tannins, saponins, minerals such as iron, calcium, potassium, magnesium, zinc and copper (1.79%–3.44%), vitamin A and C, thiamine, niacin, pyridoxine and folate (Al-Mahasneh et al., 2008; Sultan et al., 2009; Salama, 2010; Güllü and Gülcan, 2013). *N. sativa* is also rich in unsaturated and essential fatty acids and studies indicate that the volatile oil content ranges from 0.4% to 2.5% (Hosseinzadeh and Parvardeh, 2004; Sultan et al., 2009). Volatile oil contains active basic components such as thymoquinone, dithymoquinone and thymohydroquinone (Kaya et al., 2003; Güllü and Gülcan, 2013).

Table 1. The general chemical composition of *N. sativa* seeds (Ramadan, 2007)

Constituents		Chemical composition	% Range (w/w)
Oil	**Fixed oil**	Linoleic acid (Omega-6), Oleic acid, Palmitoleic acid Linolenic acid (Omega-3), Myristoleic acid, Dihomolionolenic acid, Stearic acid, Eicosadienoic acid, Myristic acid, Arachidic acid, Behanic acid, Sterols (β-sitosterol, avenasterol, stigmasterol, campesterol and lanosterol), Tocopherols (α, β, and γ) Thymoquinone, Retinol (vitamin A), Carotenoids (β-carotene)	22–38%
	Volatile oil	Thymoquinone, p-Cymene, Carvacol, α-Pinene, β-Pinene, Longifolene, t-Anethole Thymol, Thymohydroquinone, Dithymoquinone (nigellone)	0.40–1.50%
Protein		Glutamic acid, Arginine, Aspartic acid, Leucine, Glycine, Valine, Lysine, Threonine, Phenylalanine Isoleucine, Histidine, Methionine	20.8–31.2%
Carbohydrate		Glucose, Rhamnose, Xylose, Arabinose	24.9–40%
Minerals		Calcium, Phosphorus, Iron, Potassium, Sodium, Zinc, Magnesium, Manganese, Copper, Selenium	3.7–7%
Saponins		α-Hederin (melanthin), Hederagenin (melanthigenin)	0.013%
Alkaloids		Nigelicine, Nigellimine, Nigellidine	0.01%
Other Vitamins		Vitamin A, Thiamin, Riboflavin, Pyridoxine, Niacin, Folacin, Vitamin C	1–4%

The general chemical composition of *N. sativa* seeds has been presented in Table 1. Thymoquinone is the most studied of the black cumin components and researchers have aimed to clarify the mechanisms by which thymoquinone plays a role in the prevention and treatment of disease (Paarakh, 2010; Randhawa and Alghamdi, 2011; Akash et al., 2011; Ahmad and Beg, 2013). *N. sativa* can be used in various forms, as a powder, oil or extract in traditional treatment (Heshmati and Namazi, 2015; Heshmati et al., 2015). Thymoquinone was first synthesised in 1959 (Figure 3) and it was reported that thymoquinone exists as a volatile oil in a proportion of 18.4%–24.0% (Burits and Bucar, 2000; Ali and Blunden, 2003; Yüncüet al., 2013). Other analyses have indicated that the concentration of thymoquinone is 52.6 mg/100 g and 20.13 mg/100 g (Tüfek et al., 2015).

3. ANTIOXIDANT EFFECT OF *N. SATIVA*

An evaluation of the essential oil of *N. sativa* seeds, for antioxidants, showed that thymoquinone (TQ) and other components (carvacrol, *t*-anethole, and 4-terpineol) have a radical scavenging property. These constituents and the essential oil showed variable antioxidant activity when tested in the diphenylpicrylhydrazyl assay; they also effectively scavenged OH^- radical in the assay for non-enzymatic lipid peroxidation (Aruoma et al., 1997). TQ suppresed the ferric nitrilotriacetate and induced oxidative stress in Wistar rats (Khan and Sultana, 2005). Dietary *N. sativa* seeds inhibited the oxidative stress caused by oxidized corn oil in rats (Al-Othman et al., 2006). Dietary *N. sativa* (10%) neutralized the oxidative stress induced by hepatocarcinogens such as dibutylamine and sodium nitrate in albino rats by normalizing glutathione and nitric oxide levels (Gendy et al., 2007). The *N. sativa* seed oil and TQ (intraperitoneal) have protective effects on lipid peroxidation process during ischemia-reperfusion injury in rat hippocampus (Hosseinzadeh et al., 2007). Treating broiler chicks with *N. sativa* seed for 6 weeks reduced the oxidative stress in the liver by increasing the activities of myeloperoxidase, glutathione-S-transferase, catalase (CAT), adenosine deaminase, and by decreasing hepatic lipid peroxidation (Sogut et al., 2008).

The TQ pre-treatment countered the increased level of lipid peroxidation and augmented the antioxidant enzyme activities in the erythrocyte during 1,2-dimethylhydrazine-induced colon carcinogenesis in Wistar rats (Harzallah et al., 2012).

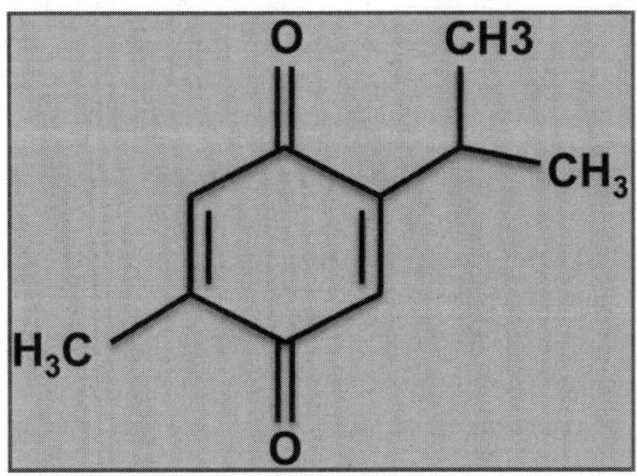

Figure 3. The molecular structure of thymoquinone (Chemical name: 2-Isopropyl-5-methylbenzo-1,4-quinone).

Thymoquinone and dithymoquinone are amongst the main antioxidant components of *N. sativa*. The intake of *N. sativa* decreases lipid peroxidation, increases antioxidant enzymes and decrease oxidative stress (Al-Mahasneh et al., 2008; Ragheb et al., 2008). It is thought that a decrease in oxidative stress renews the pancreatic betacells, maintains the integrity of betacells, increases the number and volume of islet cells, decreases insulin resistance and increases insulin secretion and aids glycation end-product inhibition (Nokoff et al., 2012). Thus, the glycaemic improvement can relieve lipid dysfunction, especially in diabetics. Furthermore, antioxidant components-reduced free radicals protect the cells against lipid peroxidation and improve the enzyme functions in lipid metabolism (Bamosa et al., 2010; Heshmati and Namazi, 2015).

3.1. Effect of *N. Sativa* on Humans and Animals Lipid Profile

Many studies on humans and animals have shown that *N. sativa* and its active component thymoquinone have positive effects by decreasing levels of serum lipids, total cholesterol (TC), triglycerides (TG) and low-density lipoproteins (LDL), whereas other studies have suggested that *N. sativa* and thymoquinone inaffective (Bamosa et al., 2002; Ragheb et al., 2008). However, no definite results have demonstrate that they increase the level of high-density lipoproteins (HDLs), which play an active role in decreasing the risk of cardiovascular disease, in particular (Razavi and Hosseinzadeh, 2014). In a study that examined the effect of *N. sativa* seed and oil on its anti-atherogenic potential in rabbits that were fed with a hypercholesteremic diet, 25 rabbits were divided into five groups. Four groups were determined as hypercholesteremic and the other group as negative normal. One of the hypercholesteremic groups was separated as a positive control group, and was fed with a diet containing 1% cholesterol for 3 weeks. During the final 8 weeks, 1 g kg^{-1} *N. sativa* powder, 0.5 g kg^{-1} *N. sativa* oil or 10 mg/day simvastatin were added, respectively, to the diet of each the groups except the positive control group. It was found that weight, plasma TC and LDL increased considerably, whereas there was no significant different in the HDL level in the group fed on a diet with 1% cholesterol, compared to the negative control group. On the other hand, plasma TC, TG and LDL levels considerably decreased in the groups whose diets contained *N. sativa* oil and seed, compared to levels in the positive group (Al-Naqeep et al., 2011). Similar to these results, intake of 10 mg ml^{-1} thymoquinone by gavage positively affected blood lipids in rats that were fed an atherogenic diet for 30 days (Ahmad and Beg, 2013). In a study conducted to examine the effects of different doses of *N. sativa* supplement on the serum lipid profile in rats, 15 rats were separated as a control group and 60 rats were fed with *N. sativa* supplement. Either100 mg, 200 mg, 400 mg and 600 mg kg^{-1} were given daily to rats for four weeks, which resulted in a significant decrease in the total cholesterol level (Kocyigit et al., 2009). In another study, individuals with type 2 diabetic patients were separated into three groups and were given *N. sativa* at a rate of 1, 2, 3 g/day for twelve weeks. Individuals who were given *N. sativa* at minimum level (1 g/day) and a maximum level (2 g/day and 3 g/day) were compared. After twelve weeks of giving *N. sativa* at 1, 2, 3 g/day rates to 2 diabetic patients, plasma TG, TC and LDL cholesterol levels were significantly lower in individuals given 2 g/day, while increasing the amount of *N. sativa* has no positive effective on individuals' lipid profiles (Kaatabi et al., 2012). In a follow-up study conducted, Latiff et al. (2014) found on premenopausal women, individuals were given either a placebo or 1,600 mg *N. sativa* powder and the lipid profile of the groups was then examined. No significant changes in the lipid profile or LDL and TG level values of the premenopausal women giving placebo compared to the initial level after 12 weeks. Additionally, no significant decrease in LDL and TG levels were found compared to the initial level in the group treated with black cumin, and the total cholesterol level was considerably lower as well as blood pressure, which is a risk factor for cardiovascular disease (Latiff et al., 2014).

In a study in which menopausal women were given 1 g/day *N. sativa* powder after breakfast for two months, the weight of women decreased compared with that of the control group, although the difference was not statistically significant; however, their TC, TG, LDL and HDL levels considerably improved (Ibrahim et al., 2014). Therefore, it has been claimed that intake of *N. sativa* in different forms can be used as a supportive for drugs that decrease the lipid profile (Razavi and Hosseinzadeh, 2014).

4. Anticancer Effects of *N. Sativa*

Cancer is considered to be one of the widespread health concerns since it killed half a million of people in America during 2013 and 8.2 million people worldwide (WHO, 2015). *N. sativa* and its active component thymoquinone are claimed to exhibit anticancer activity by causing the death of cancer cells or by preventing genetic changes in normal cells (Shafiq et al., 2014).

Many studies have shown that *N. sativa* and thymoquinone have antioxidant properties, and that they increase the activities of antioxidant enzymes such as superoxide dismutase (SOD), catalase (CAT) and glutathione peroxidase (Bordoni et al., 2019), therefore, the positive effects of *N. sativa* against cancer types possibly occurs via antioxidant effects (Badary et al., 2003; Khader et al., 2010; Randhawa and Alghamdi, 2011). *N. sativa* has been shown to kill various cancerous cell types, and to increase macrophage cell number and activation (Shoieb et al., 2003; El-Mahdy et al., 2005; Mbarek et al., 2007; Chehl et al., 2009; Yüncü et al., 2013; Darakhshan et al., 2015).

4.1. Mechanism for Thymoquinone Action in Cancer Therapeutics

For thousands of years, black cumin has been used in different traditional systems of medicine, and thymoquinone, dithymoquinone, thymohydroquinone and thymol are the active ingredients found in black cumin seeds. Several studies have indicated that thymoquinone has notable activities in the prevention and treatment of various types of diseases including cancers by targeting different biochemical, molecular and physiological pathways (Khan et al., 2017). It has been evident that thymoquinone can inhibit DNA synthesis and proliferation in cancer cells by interfering with the structure of DNA (Salem et al., 2015). In pathogenesis of different cancers, oxidative stress has vital roles. Usually the antioxidant systems prevent oxidative DNA damage by scavenging free radicals, and the antioxidant activity of thymoquinone to protect cellular damage is well documented (Nagi and Mansour, 2000). Thymoquinone can upregulate proapoptotic genes and proteins, such as Bax/Bak, or downregulate antiapoptotic genes and proteins, such as Bcl-2, Bcl-xL, among others, as well as modulating the caspasepathway, and thus controlling cancer cell growth by inducing apoptosis (Khan et al., 2011).

It is also evident that thymoquinone induces apoptosis by interfering with the activity and expression of various other molecular targets, such as peroxisome proliferator-activated receptor(PPAR)-γ, phosphatase and tensin homolog (PTEN) and signal transducer and activator of transcription (STAT)3, and through the generation of reactive oxygen species (ROS) (Woo et al., 2012). Certain tumor suppressor genes and proteins have been found to be overexpressed or activated by thymoquinone; for example breast cancer type 1 susceptibility protein (BRCA1), among others, whereas certain oncogenic signaling molecules and pathways, like phosphoinositide 3 kinase (PI3K)/Akt and mitogen-activated protein kinase (MAPK)/ERK, have been found to be inhibited by thymoquinone (Şakalar et al., 2016).

4.2. Lung Cancer

Mabrouk et al. (2002) showed that supplementation of diet with honey and *N. sativa* has a protective effect against MNU (methylnitrosourea)-induced oxidative stress, inflammatory response and carcinogenesis in lung, skin and colon. Also, Swamy and Huat (2003) mentioned the antitumor activity of α-hederin from *N. sativa* against LL/2 (Lewis Lung carcinoma) in BDF1 mice. However, Rooney and Ryan (2005) reported that α-hederin and TQ, the two principal bioactive constituents of *N. sativa* enhance neither cytotoxicity nor apoptosis in A549 (lung carcinoma), HEp-2 (larynx epidermoid carcinoma) cells. According to data from the American Cancer Association in 2012, lung cancer caused approximately 20% of total cancer deaths and this rate reached 27% in 2015 (American Cancer Society, 2015). *N. sativa* supplement and *N. sativa* seed extract demonstrated cytotoxicity properties against lung sarcoma cells (Rooney and Ryan, 2005). In particular, thymoquinone has anticancer properties against lung cancer cell numbers and prevented cell proliferation by approximately 90% (Shafiq et al., 2014).

4.3. Breast Cancer

The breast cancer considers as the primary and secondary causes of the females death in undeveloped and developed countries, respectively (Shafiq et al., 2014). The liquid and alcoholic extracts of *N. sativa* have been shown to be effective in inactivating a breast cancer cell line (MCF-7) and have an effect on life span. Therefore, these extracts are proposed to represent a promising treatment for breast cancer (Farah and Begum, 2002; Shafiq et al., 2014).

4.4. Colon Cancer

Colon cancer causes the second-highest number of deaths among different cancers and 690,000 per annum die from it (American Cancer Society, 2015). It has been suggested that *N. sativa* can reduce DNA damage and prevent carcinogenesis in colon tissues exposed to toxic agents (de Souza Grinevicius et al., 2016). A study that analysed the relationship between *N. sativa* and colon cancer demonstrated that thymoquinone inhibits the formation of 5-lipoxygenase products such as 5-hydroxeicosa-tetraenoic acids, which are necessary for colon cancer cells (El-Mahmoudy et al., 2002). It has been shown that the effect of thymoquinone depends on the type of colon cancer cell since it affects HCT-116 colon cancer cells by increasing apoptosis, but has no effect on HT-29 colon cancer cells (Rooney and Ryan, 2005; Abukhader, 2012). In addition, Salim and Fukushima (2003) demonstrated that the volatile oil of *N. sativa* has the ability to inhibit colon carcinogenesis of rats in the post-initiation stage, with no evident adverse side effects. Norwood et al. (2006) suggested TQ as chemotherapeutic agent on SW-626 colon cancer cells, in potency, which is similar to 5-flurouracil in action. However, on HT-29 (colon adenocarcinoma) cell, no effect of TQ was found (Rooney and Ryan, 2005). Thymoquinone administered to mice reduced the incidence of stomach tumors (Badary et al., 1999) via mechanisms that included inhibition of DNA synthesis and the promotion of apoptosis by inhibiting cell growth in the G1 phase, and it also inhibited cellular proliferation and induced apoptosis in colon cancer cells by triggering a p53-dependent mechanism (Gali-Muhtasib et al., 2004).

4.5. Blood Cancer

El-Mahdy et al. (2005) reported that TQ exhibits anti-proliferative effect in human myeloblastic leukemia HL-60 cells. Derivatives of TQ bearing terpene-terminated 6-alkyl residues were tested in HL-60 cells and 518A2 melanoma by Effenberger et al. (2010) and they found the derivatives induce apoptosis associated with DNA laddering, decrease mitochondrial membrane potential and cause a slight increase in reactive oxygen species. Swamy and Huat (2003) observed that α-hederin also induced death of murine leukemia P388 cells by a dose- and time-dependent increase in apoptosis.

4.6. Pancreatic Cancer

Chehl et al. (2009) showed that TQ induced apoptosis and inhibited proliferation in pancreatic ductal adenocarcinoma (PDA) cells. They also suggested TQ as a novel inhibitor of pro-inflammatory pathways, which provides a promising strategy that combines anti-inflammatory and proapoptotic modes of action. TQ also can abrogate gemcitabine- or oxaliplatin-induced activation of NF-kappa B, resulting in the chemosensitization of pancreatic tumors to conventional therapeutics (Banerjee et al., 2009). The high molecular weight glycoprotein mucin 4 (MUC4) is aberrantly expressed in pancreatic cancer and contributed to the regulation of differentiation, proliferation, metastasis, and the chemoresistance of pancreatic cancer cells. Torres et al. (2010) evaluated the down-regulatory effect of TQ on MUC4 in pancreatic cancer cells. Finally, Rooney and Ryan (2005) did not find any preventive role of TQ on MIA PaCa-2 (pancreas carcinoma) cells.

4.7. Hepatic Cancer

The cytotoxic activity of *N. sativa* seed was tested on the human hepatoma HepG2 cell line by Thabrew et al. (2005) and 88% inhibitory effect on HepG2 was found after 24 h. incubation with different concentrations (0-50 mg ml^{-1}) of the *N. sativa* extract. Nagi and Almakki (2009) reported that oral administration of TQ is effective in increasing the activities of quinine reductase and glutathione transferase and makes TQ a promising prophylactic agent against chemical carcinogenesis and toxicity in hepatic cancer.

4.8. Skin Cancer

Topical application of *N. sativa* extract inhibited two-stage initiation/promotion [dimethylbenz[a]anthracene (DMBA)/croton oil] skin carcinogenesis in mice. Again, intraperitoneal administration of *N. sativa* (100 mg kg^{-1} body wt) 30 days after subcutaneous administration of MCA (20-methylcholanthrene) restricted soft tissue sarcomas to 33.3% compared with 100% in MCA-treated controls (Salomi et al., 1991).

4.9. Fibrosarcoma

TQ from *N. sativa* was administrated (0.01% in drinking water) one week before and after MCA (Methyl Chol Anthrene) treatment significantly inhibited the tumor incidence (fibrosarcoma) and tumor burden by 43% and 34%, respectively, compared with the results in the group receiving MCA alone (Salomi et al., 1991). Moreover, TQ delayed the onset of MCA-induced fibrosarcoma tumors. Also *in vitro* studies showed that TQ inhibited the survival of fibrosarcoma cells with IC50 of 15 mM (Badary and Gamal, 2001). Oil of *N. sativa* also decreased the fibrinolytic potential of the human fibrosarcoma cell line (HT1080) *in vitro* (Awad, 2005).

4.10. Renal Cancer

Khan and Sultana (2005) reported the chemo-preventive effect of *N. sativa* against ferric nitrilotriacetate (Fe-NTA) induced renal oxidative stress, hyper-proliferative response and renal carcinogenesis. Treatment of rats orally with *N. sativa* (50 - 100 mg kg^{-1} body wt) resulted in significant decrease in H_2O_2 generation, DNA synthesis and incidence of tumors (Khan and Sultana 2005).

4.11. Prostate Cancer

TQ from *N. sativa* inhibited DNA synthesis, proliferation and viability of cancerous (LNCaP, C4-B, DU145, and PC- 3), but not non-cancerous (BPH-1) prostate epithelial cells by down-regulating AR (androgen receptor) and E2F-1 (a transcription factor) (Kaseb et al., 2007). In this study, they suggested TQ as effective in treating hormone-sensitive as well as hormonerefractory prostate cancer. Also, Yi et al. (2008) found that TQ blocked angiogenesis *in vitro* and *in vivo*, prevented tumor angiogenesis in a xenograft human prostate cancer (PC3) model in mouse and inhibited human prostate tumor growth at low dosage with almost no chemotoxic side effects. Furthermore, they observed that endothelial cells were more sensitive to TQ-induced cell apoptosis, cell proliferation and migration inhibition compared with PC3 cancer cells. TQ also inhibited vascular endothelial growth factor-induced extracellular signal-regulated kinase activation but showed no inhibitory effects on vascular endothelial growth factor receptor 2 activation (Tingfang et al., 2008).

4.12. Cervical Cancer

Shafi et al. (2009) reported that methanol, n-Hexane and chloroform extracts of *N. sativa* effectively killed HeLa (human epithelial cervical cancer) cells by inducing apoptosis. Effenberger et al. (2010) tested terpene-terminated 6-alkyl residues of TQ on multidrug-resistant KB-V1/Vb1 cervical carcinoma and found the derivatives inducing cell death by apoptosis.

4.13. Molecular Mechanisms of *N. Sativa* Action against Cancer

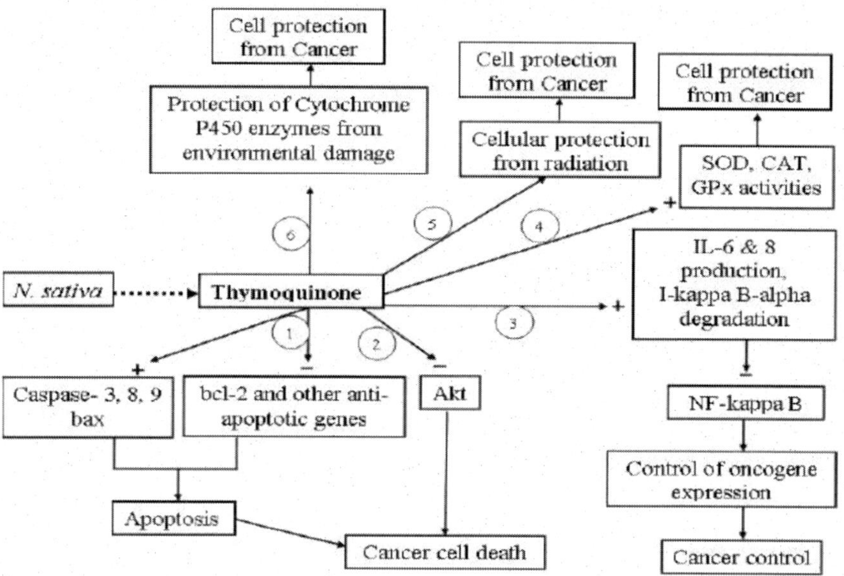

'+' indicates increasing effect and '-' indicates decreasing effect.

Figure 4. Possible mechanisms of thymoquinone (TQ) action against cancer. (1) TQ induces apoptotic cell death in cancerous tissues by upregulating expression of apoptotic genes (caspases and bax) and down-regulating expression of anti-apoptotic genes (e.g., bcl 2); (2) TQ suppresses Akt activation by dephosphorylation and thus blocks cancer cell survival; (3) TQ deactivates NFkappa B pathway by inducing cytokine production, and thus control oncogenic expression; (4) TQ increases the activities of antioxidant enzymes and protects cell against cancer; (5) TQ protects normal cells' injury caused by ionizing radiation in the treatment of cancer; (6) TQ prevents CYP450 enzymes from damage.

Cancers are the abnormal cell growth caused by genetic alteration. The anti-cancer agent defined as the agent that can protect genetic material from alteration or kill the genetically altered cancer cells. The active ingredients (mainly TQ) from *N. sativa* kills cancer cells by several molecular pathways (Figure 4).

5. ANTIMICROBIAL EFFECT OF *N SATIVA*

5.1. Antibacterial Effect of *N Sativa*

New antimicrobial agents are intensively investigated due to pathogenic bacterial infections and microbial resistance, which have become a major health problem worldwide leading to increasing use of medicinal plants (Emeka et al., 2015). Therefore, many studies discussed antibacterial efficacy of black seeds. Thymoquinone exhibited a significant bactericidal activity against the majority of the tested bacteria (MICs values ranged from 8 to 32 μg/ml) especially Gram positive cocci (*Staphylococcus aureus* ATCC 25923 and *Staphylococcus epidermidis* CIP 106510). Crystal violet assay demonstrated that the minimum biofilm inhibition concentration (BIC50) was reached with 22 and 60 μg/ml for *Staphylococcus aureus* ATCC 25923 and *Staphylococcus epidermidis* CIP 106510 respectively (Chaieb et al., 2011). Black seeds have been found effective against Helicobacter *pylori* compared with triple therapy (Salem et al., 2010). Black seed extract has also been found to have several multidrug resistant clinical bacterial effects and the inhibition of the bacterial growth was due to the presence of thymoquinone and melanin (Islam et al., 2013).

N. sativa essential oil (included *p*-cymene, TQ, α-thujene, longifolene, β-pinene, α-pinene and carvacrol) exhibited different biological activities including antifungal, antibacterial and antioxidant potentials. *N. sativa* essential oil completely inhibited different Gram-negative and Gram-positive bacteria (Morsi, 2000). *N. sativa* oil also exhibited stronger radical scavenging activity against DPPH radical in comparison with synthetic antioxidants (Bordoni et al., 2019). The mechanism of the antimicrobial effect of *N. sativa* seeds has not been reported, its antimicrobial property could be attributed to the active constituents particularly TQ and melanin

(Bakathir and Abbas, 2011). Their broad spectrum of activity may be the reason of that the key processes of the organisms are affected (Monika et al., 2013). Concentration dependent inhibition of Gram-positive; *Staphylococcus aureus* and Gram-negative; *Pseudomonas aeruginosa*, *Escherichia coli* and pathogenic yeast *Candida albicans* in filter paper discs impregnated with ethyl ether extract of *N. sativa* (25-400 micrograms/disc) was observed, the extract showed antibacterial synergism with streptomycin and gentamicin and exhibited additive antibacterial action with doxycycline, spectinomycin, erythromycin, tobramycin, ampicillin, chloram-phenicol, nalidixic acid, lincomycin and sulfa-methoxazole trimethoprim combination (Hanafy and Hatem, 1991).

To investigate the antibacterial effect of crude extracts of *N. sativa*, various bacterial isolates which included of 16 Gram-negative and 6 Gram-positive, these isolates, particularly Gram-negative bacteria, showed multiple resistance against antibiotics. Crude alkaloid and water extracts were the most effective extracts, and especially against Gram-negative isolates they were effective (Dorman and Deans, 2000). In another study, a population of 7.0 log CFU of each strain of *Listeria monocytogenes* was inoculated on duplicate plates having antibiotic medium one agar. Each discs (6 mm diameter), impregnated with 10 μl of black seed oil, or gentamicin (positive control). *N. sativa* seed oil had a strong antibacterial activity against all the strains of *L. monocytogenes*, yielding a significantly greater inhibition zone than that of gentamicin (positive control) ($P < 0.01$). The mean zones of inhibition produced by *N. sativa* seed oil and gentamicin were 31.50 ± 1.0 and 14.80 ± 0.50 mm, respectively (Nair et al., 2005). All tested strains of Methicillin resistant *S. aureus* (MRSA), which is one of the commonest pathogens encountered inlaboratory and in clinic, were sensitive to ethanolic extract of *N. sativa* seeds at a concentration of 4 mg/discs, and the extract had a minimum inhibitory concentration (MIC) range of 0.2-0.5 mg ml^{-1} (Hannan et al., 2008).

The two main components of black seed essential oil, thymoquinone (TQ) and thymohydroquinone (THQ) were investigated for their antibacterial activity against *Escherichia coli*, *Pseudomonas aeruginosa*, *Shigella flexneri*, *Salmonella Typhimurium*, *Salmonella Enteritidis* and

Staphylococcus aureus. Both TQ and THQ exerted antibacterial activity against gram-positive and gram-negative bacteria regardless to their susceptibility to antibiotics. *S. aureus* was highly susceptible to TQ, since, three and 6 μg ml^{-1} were enough to inhibit and kill the bacteria respectively. On the contrary the concentration of THQ required to inhibit and kill S. aureus was 400 and 800 μg ml^{-1} respectively which is 100 times more than that of TQ. Gram-negative bacteria were less susceptible to both TQ and THQ and their MIC and MBC ranged between 200 and 1600 μg ml^{-1} (Halawani, 2009).

In MIC determination, TQ was active against all the strains that studied. The main antibacterial activity was seen against *Streptococcus constellatus* (MIC 4 μg ml^{-1}). The essential oil showed the strongest activity against *Streptococcus mitis*, *S. mutans*, *S. constellatus* and *Gemella haemolysans* (MIC 2.13 mg ml^{-1}), but it was ineffective against *Enterococcus faecalis* and *Enterococcus faecium* (MIC > 8.5 mg ml^{-1}) (Harzallah et al., 2011). By the disc diffusion method, TQ (150μg/disk) was effective especially against *S. mutans* and *S. mitis* (zone of inhibition were: 24.5 ± 0.71 and 22 ± 1.41 mm, respectively), while it showed a weak antibacterial activity against *E. faecalis*, *E. faecium* and *Streptococcus salivarius* (9 ± 0.00, 9.5 ± 0.71 and 9.5 ± 0.71 mm, respectively) (Harzallah et al., 2011). Although the essential oil (2.43 mg/disc) have high activity against *S. mitis*, *S. oralis*, *S. mutans*, *S. constellatus* and *G. haemolysans* with a zone of inhibition ranged from 13.5 to 15.5 mm, besides, it was ineffective against *E. faecalis*, *E. faecium* and *S. salivarius* (Harzallah et al., 2011).

In addition, the growth inhibition of *S. aureus* was seen at concentration of 300 mg ml^{-1} of *N. sativa* seeds compared with control, and confirmed with azithromycin as positive control which may be due to TQ (Bakathir and Abbas, 2011). Moreover, the aqueous extract of *N. sativa* seed showed less antibacterial effect compared to the methanol extract. At concentration of 20 mg ml^{-1} the methanol extract of seed was effective against *S. pyogenes* (10 mm zone of inhibition) but at 100 mg ml^{-1} was effective against *P. aeruginosa* (20 mm zone of inhibition), *S. pyogenes* (15 mm zone of inhibition) and at concentration of 50 mg ml^{-1} exhibited modest effective against *S. pyogenes* (10 mm zone of inhibition), *Klebsiella pneumoniae* (11 mm zone of inhibition), and *Proteus vulgaris* (12 mm zone of inhibition) (Hasan et al., 2013). *N. sativa* seed extract has been loaded into the polymeric micelle and its effectiveness has been evaluated against Gram-positive strain of *S. aureus*, *Bacillus subtilis* and a Gram-negative *E. coli* (Deepak et al., 2011).

5.2. Antiviral Effect of *N Sativa*

Apoptosis is caused by viral infections leading to lymphocyte depletion in the host cell and antioxidants can inhibit apoptosis which induced by viruses in addition to inhibit the viral replication in target cells, so antiviral and antioxidant effects can be linked together (Peterhans, 1997). To investigate the antiviral effect of *N. sativa* oil, murine cytomegalovirus (MCMV) as a model was used. Intraperitoneal administration of *N. sativa* oil to mice completely inhibited the virus titers in spleen and liver on day 3 of infection. Viral load in the liver and spleen of the control had a high difference with *N. sativa* oil treated mice, 45 × 104 vs. 7 × 104 and 23 × 103 vs. 3 × 103, respectively. This antiviral effect accorded with raising the serum level of interferon-gamma and increased numbers of helper T cells ($CD4^+$), suppressor function and numbers of macrophages. On the tenth day of infection, the virus titer was undetectable in spleen and liver of *N. sativa* oil treated mice, whereas it was detectable in control mice, so the *in vivo* treatment with *N. sativa* oil induced a remarkable antiviral effect against MCMV infection (Salem and Hossain, 2000). By nonspecific cells including natural killer cell (NK cells), and specific cells including CD4 and CD8 T cells, immunity produced to viral infection is controlled. The antiviral effect of the *N. sativa* oil is related with increasing response of CD4 cells (Salem and Hossain, 2000). In a research, the patient with hepatitis C virus (HCV) infection, who was not ineligible for IFN-α therapy, had received the capsule of *N. sativa* oil (450 mg) for three successive months, for three times in a day and the findings showed a significant decreased of the viral load and improvement of the oxidative stress due to augmented total antioxidant activity, total protein and albumin, and improved Red Blood Cell (RBC) and platelet counts in HCV patients. The augmented RBC count can attributed to the lowering of the membrane lipid peroxide level, leads to reduce the incidence of hemolysis (Barakat et al., 2013). Diminish the blood glucose levels, implying that it may provide a potential modulatory influence on HCV induced glucose intolerance, improve in the lower limb edema also was seen (Barakat et al., 2013).

5.3. Antifungal Effect of *N Sativa*

N. sativa oil showed antifungal activity against most pathogenic fungi (*Candida tropicalis*, *Aspergillus flavus*) (Gupta et al., 2012). Thymoquinone, which are the main composition of the oil, showed antifungal activity against most fungal strains (Rogozhin et al., 2011; Gupta et al., 2012). Thymoquinone has antifungal activity against *Cryptococcus albidus* which is 20.83 μg ml^{-1}, *Candida albicans* 23.33 μg ml^{-1}, *Issatchenkia orientalis* 25.0 μg ml^{-1} and *Aspergillus fumigatus* 23.40 μg ml^{-1} (Suthar et al., 2010). In another study of *N. sativa* antifungal effect on dermatophyte fungal strains, it was found that the essential oil extract of thymoquinone has an effective antifungal activity on Microsporum canis, Trichophyton mentagrophytes, and Microsporum gypseum (Mahmoudvand et al., 2014). In addition, Khosravi et al. (2011) concluded that *Cuminum* cyminum, *Ziziphora* clinopodioides and *N. sativa* oils possess antifungal activities to inhibit the growth of Aspergillus *fumigatus* and Aspergillus *flavus*. The antifungal activity of the oils was evident at the morphological level. Due to the antifungal activity of these oils and their availability as natural volatile products, they might be of use in future studies of antifungal agents (Khosravi et al. 2011). Researchers who studied treatment of fungal infections by using natural products found that *N. sativa* has an enhancing antifungal effect (Bita et al., 2012; Asdadi et al., 2014). Another research that depended onmicro well dilution assay was conducted against three human pathogenic fungal strains *A. flavus*,

Aspergillus niger, and *Candida albicans* (Javed et al., 2012). Moreover, the extracts of essential *N. sativa* oil showed effective antifungal activity against Candida *albicans*, Candida *tropicalis* and Candida *krusei* at MIC values of 16–64 μg ml^{-1} (Piras et al., 2013). Moreoever, Harzallah et al. (2012) stated that Tunisian *N. sativa* fixed oil showed a validation for the folk use of this oil as an antibacterial and antifungal medicine. Therefore, the existence of various chemical compounds in *N. sativa* and their mechanism of action make it a good candidate as an antifungal agent.

A study was done to investigate the antidermatophyte effects of ether extract of *N. sativa* seeds and TQ. In this research, the test was performed by using the agar diffusion method with serial dilutions of ether extract of *N. sativa* seed, TQ and griseofulvin. The species were used consists of eight species of dermatophytes, four species of *Trichophyton rubrum* and one each of *Epidermo-phyton floccosum Trichophyton interdigitale, Microsporum canis, and Trichophyton menta-grophytes.* The results showed that the MICs of the ether extract of *N. sativa* seed and TQ were in range of 10-40 and 0.125-0.250 mg ml^{-1}, respectively, whereas for griseofulvin was between 0.00095 to 0.01550 mg ml^{-1} (Aljabre et al., 2005). Two novel defensins named Ns-D1 and Ns-D2 from seeds of *N. sativa*, were isolated, purified and sequenced. The Ns-D1 and Ns-D2 peptides exhibited strong divergent antifungal activity against a number of phytopathogenic fungi. The mechanism is due to expected interaction of defensins with specific sphingolipids on the fungal membranes (Rogozhin et al., 2011). The growth of *Aspergillus parasiticus* (CBS 921.7) and *Aspergillus flavus* (SQU 21) strains and also the production of aflatoxin B1 by this fungus, were evaluated. The inhibition of aflatoxin B1 production by *A. flavus* and *A. parasiticus* with different concentrations of *N. sativa* oil (1, 2 and 3 ml/100 ml) were in the range of 49.7-58.3% and 32-48%, respectively, but different concentrations of *N. sativa* oil displayed insignificant effect on the growth of other *Aspergillus* species. *N. sativa* oil might have metabolic effects on biosynthesis pathways of aflatoxin (El-Nagerabia et al., 2012). *N. sativa* essential oil represented a significant antidermatophytic activity, that the maximum zone of inhibition was seen in *Microsporum gypseum* with inhibition zone of (IZ) 38 mm and activity index (AI): 1.90 (Sunita and Meenakshi, 2013).

5.4. Antiparasitic Effect of *N Sativa*

N. sativa seeds methanolic extractions were tested on *Plasmodium yoelii* infection to see its efficacy. It was found that *N. sativa* methanolic extraction showed 94%, $P < 0.05$, which showed an excellent suppression compared with chloroquine, which is the drug of choice for *Plasmodium yoelii* infection treatment (methanolic extract of the drug led to 86%), and such antimalarial activity was because of antioxidant effect of the extract on *Plasmodium* infected mice. The study was improved to see the antioxidative status in red blood cells, and hepatocytes of infected mice were seen (Tembhurne et al., 2014; Adam et al., 2016).

N. sativa oil possesses other activities against cestodes and nematodes action (Mahmoud et al., 2002). Also, *N. sativa* oil had an excellent effect in minimizing the total number of *Schistosoma mansoni* worms in liver and reducing the total number of ova that was found in both liver and intestine (Forouzanfar et al., 2014). In addition, El Shenawy et al. (2008) noted that the treatment of *Schistosoma mansoni*-infected mice with *N. sativa* oil had an improvement Hematological and biochemical parameters including liver and kidney functions were measured to assess the progress of anemia, and the possibility of the tissue damage. Serum total protein level, albumin, globulin and cholesterol, Malondialdehyde (MDA) and glutathione (GSH) levels in the liver tissues in schistosomiasismice compared to the infected untreated ones. *N. sativa* oil and *thymoquinone* was used to test their efficacy against *Sch. mansoni* on infected mice. Results showed a decrease in chromosomal abnormalities, especially on chromosomes 2 and 6, and some in chromosomes 13 and 14 when *N. sativa* oil and *thymoquinone* are used in treatment compared with control group (Aboul-Ela, 2002).

Another study carried by Mohamed et al. (2005) achived significant results as a strong effect against all used parasites (*Sch. mansoni, miracidia, cercariae*, and adult worms) and even on their eggs. Also *N.*

sativa seeds possed an oxidative activity against adult worms which decrease the activities of some enzymes such as glutathione reductase, antioxidant enzymes, and enzymes of glucose metabolism. The presence of antioxidant compounds in *N. sativa* may lead to the collection of free radicals and inactivation of them, which may propose significant marketing advantage, due to consumer preference for antioxidant rich products.

6. Effects of *Nigella Sativa* on Health and Pharmacological Uses

A wide range of studies have been conducted concerning the biological activities and curative properties of black cumin (Ahmad and Beg, 2013). *N. sativa* is used in the treatment of many diseases in many countries globally. Its beneficial effects on health, especially against diseases such as cancer, diabetes and cardiovascular disease, have been highlighted (Leong et al., 2013; Shafiq et al., 2014; Singh et al., 2014; Entok et al., 2014; Bamosa, 2015). The summary of possible mechanisms of action of *N. sativa* in certain diseases is shown in Figure 5.

6.1. Antihyperlipidemic and Antihypercholesteremic Effects of *N. Sativa*

Hypercholesterolemia is reflected by an increase in triglyceride (TG), total cholesterol (TC), low-density lipoprotein (LDL) (bad chlolesterol), HDL and very low-density lipoprotein (VLDL) levels (Ahmad and Beg, 2013). An increase in the HDL level and a decrease in the LDL level in the circulation have a positive effect on the risk reducing of cardiovascular disease (Mani and Rohatgi, 2015). *N. sativa* and its important active component thymoquinone demonstrate an antihypercholesteremic effect by decreasing the level of 3-hydroxy-3-methyl-glutaryl-CoA (HMG-CoA) reductase enzyme, which is the rate limiting enzyme in cholesterol synthesis, to cause protective effects on dyslipidemia. Moreover, *N. sativa* is reported to display antihyperlipidemic properties by stimulating paraoxonase enzyme (PON1), which functions as an antioxidant, due to its LDL protective property against oxidation and its ability to neutralize radicals including hydrogen peroxide, to increase the activity of arylesterase, the protein indicator of the PON1 enzyme (Türkoğlu et al., 2008; Ahmad and Beg, 2013).

Nigella sativa **(Thymoquinone)**	
Mechanism of action	**Disease**
HMG-COA ↓ Reductase activity arylesterase ↑ Antioxidant properties shown in cholesterol metabolism	Dyslipidaemia Metabolic syndrome
Reducing oxidative stress Blocking calcium channels Diuretic effect Hypotension (soothing heartbeat) effects	Hypertension
Increasing insulin sensitivity Blocking gluconeogenesis Decreasing glucose absorption Increasing insulin secretion Increasing β-cell proliferation	Hyperglycaemia Type 2 DM
Dietary intake ↓	Obesity
Increasing the activity of antioxidant enzymes such as superoxide dismutase (SOD), catalase (CAT) and glutathione peroxidase (GPX) Protection against cancer cells Inhibiting cancer cell growth Various types of cancer-cell death Macrophage cell number and activation ↑ Inhibiting metastasis	Cancer

Figure 5. Possible effect mechanism of *N. sativa* at certain disease.

The positive effects of *N. sativa* and its essential active component thymoquinone on cholesterol stem from their regulatory roles in antioxidant and gene metabolism. Their antioxidant properties are particularly important for the prevention of free radical formation due to diets with a high level of saturated fat and cholesterol, and for the prevention of oxidative stress and hypocholesteremia, because *N. sativa* is stated to have a protective role especially in LDL oxidation (Türkoğlu et al., 2008; Ahmad and Beg, 2013).

6.2. Antihypertensive Effects of *N. Sativa*

Another important risk factor for cardiovascular disease is hypertension. Arabs have used *N. sativa* seed together with honey or garlic for the treatment of hypertension in traditional medicine. It has been suggested that *N. sativa* extract reduces blood pressure in dogs. It has also been claimed that another antihypertensive effect of *N. sativa* oil might result from its diuretic effect (Salama, 2010).

An increase in oxidative stress is associated with the pathogenesis of hypertension. Blood pressure increases, depending on the imbalance between the antioxidant defence mechanism and free-radical production. An excessive increase in reactive oxygen products reduces the bioavailability of nitric oxide in endothelium dysfunction and increases the total peripheral resistance (Leong et al., 2013). In human and animal studies, *N. sativa* and its active component thymoquinone is reported to contribute to a reduction in blood pressure and to reduce hypertension via various mechanisms, such as antioxidant properties, calcium-channel blockage, and diuretic and hypotensive (soothing heartbeat) functions (Ahmad et al., 2013; Keyhanmanesh et al., 2014).

In a study conducted on 70 healthy individuals with an age range of 34–63 years, a body weight range of 55–75 kg, a systolic blood pressure range of 110–140 mm Hg and a diastolic blood pressure range of 60–90 mm Hg, individuals were divided into two groups – control and intervention. The intervention group was provided with 2.5 mL *N. sativa* oil after meals, every 12 hours twice a day (5 mL/day total) for eight weeks. A significant decrease in diastolic and systolic blood pressure resulted in the group that was given *N. sativa* oil (Fallah Huseini et al., 2013). In another randomized controlled double-blind dose-response study that lasted 8 weeks, on 119 men between 35 and 50 years with mild hypertension, individuals were divided into three groups and were given a placebo, 100 mg or 200 mg *N. sativa* extract A significant decrease in the systolic and diastolic blood pressure was observed in the intervention group compared to initial levels and to those in individuals with the placebo; furthermore, extract usage decreased the diastolic and systolic blood pressure, depending on its dose. These results indicate that use of *N. sativa* extract for two months has a positive effect on lowering blood pressure in individuals with mild hypertension (Dehkordi and Kamkhah, 2008).

6.3. Anti-Diabetic Effect of *N. Sativa*

N. sativa and its active component thymoquinone have been shown to have positive effects in controlling glucose levels and lipid profiles in diabetics (Heshmati and Namazi, 2015). Although the molecular mechanism of thymoquinone on insulin secretion has not been completely clarified, it is reported that thymoquinone causes an increase in glucose use by increasing the serum concentration and decreasing high levels serum glucose, and decreasing blood glucose by preventing gluconeogenesis (Benhaddou-Andaloussi et al., 2008; Heshmati and Namazi, 2015; Kaatabi et al., 2015).

N. sativa seeds are traditionally used in the management of diabetes mellitus in indigenous systems of medicine and folk remedies. Defatted extract of *N. sativa* seed is reported to increase glucose induced insulin release from isolated rat pancreatic islets *in vitro* (Rchid et al., 2004). The effect of *N. sativa* extracts have been investigated on insulin secretion *in vitro* in rat pancreatic islets in the presence of glucose (8.3 mmol L^{-1}), and the results showed that the antidiabetic properties of *N. sativa* seeds may partially be mediated by the stimulation of insulin release, especially by the basic subfraction of the seed. The possible insulinotropic property of *N. sativa* oil has also been studied in Streptozotocin or streptozocin (STZ) plus nicotinamide (NA) induced diabetes mellitus in hamsters. *N. sativa* oil treatment for four weeks decreased blood glucose and increased serum insulin (Fararh et al., 2002). Immunohistochemical staining revealed the presence of insulin in the pancreas from *N. sativa* oil treated group, suggesting that the hypoglycemic effect results, at least partly, from a stimulatory effect on β-cell function. The oil of *N. sativa* significantly lowered blood glucose in STZ diabetic rats after 2, 4, and 6 weeks (El-Dakhakhny et al., 2002). A study of the effect of *N. sativa* oil on insulin secretion from isolated rat pancreatic islets in the presence of glucose indicated that its hypoglycemic effect might be mediated by extra-pancreatic actions rather than by stimulation of insulin release. The hypoglycemic effect of *N. sativa* oil (400 mg kg^{-1}) is partly due to decreased hepatic gluconeogenesis (Fararh et al., 2004). Indazole-type alkaloid 17-O-(β-D-glucopyranosyl)-4-O-methyl nigellidine present in the defatted extract of *N. sativa* seeds increased glucose consumption by hepatocytes (HepG2 cells) *in vitro* through activation of AMP activated protein kinase (AMPK) (Yuan et al., 2014).

N. sativa extract given orally for 2 months not only decreased lipid peroxidation and increased antioxidant defence system, but also prevented the lipid peroxidation-induced liver damage in diabetic rabbits (Meral et al., 2001). Daily oral administration of ethanol extract of *N. sativa* seeds (300 mg kg^{-1}) to STZ-diabetic rats for 30 days reduced the elevated levels of blood glucose, lipids, plasma insulin, and improved altered levels of lipid peroxidation products and antioxidant enzymes in liver and kidney (Kaleem et al., 2006). This suggested that in addition to antidiabetic activity, *N. sativa* seeds may control diabetic complications through antioxidant effects. Treatment of *N. sativa* oil (0.2 ml kg^{-1}) for 30 days decreased the elevation in serum glucose and restored lowered serum insulin with partial regeneration or proliferation of pancreatic β-cells in STZ-diabetic rats (Kanter et al., 2003). The possible protective effects of *N. sativa* (0.2 ml kg^{-1}) against β-cell damage from STZ-diabetes in rats have been evidenced by the observed decrease in lipid peroxidation and serum nitric oxide, and increase in the activities of antioxidant enzymes in the pancreas (Kanter et al., 2004). Increased staining for insulin and preservation of β-cell numbers were evident in *N. sativa*–treated diabetic rats. This suggests that *N. sativa* treatment exerts a protective effect on diabetes by decreasing oxidative stress and preserving pancreatic β-cell integrity. *Nigella sativa* oil administration (daily) to STZ-induced diabetic rats maintained on a high-fat diet significantly induced the gene expression of insulin receptor (Balbaa et al., 2016). *N. sativa* oil upregulated the expression of insulin-like growth factor-1 (IGF-1) and phosphoinositide-3 kinase, whereas the expression of ADAM metallopeptidase domain 17 (ADAM-17) was downregulated. Also, the *N. sativa* oil significantly reduced blood glucose level, individual lipid profile, oxidative stress markers, serum insulin or insulin receptor ratio and the TNF-α, confirming that *N. sativa* oil has an antidiabetic activity. Thus, the daily *N. sativa* oil treatment improves insulin-induced signalling.

Hyperglycaemia is an important risk factor for the development and progression of the macrovascular and microvascular complications that occur in diabetes. The expression of apoptotic markers in the medial aortic layer of diabetic rats and the effects of *N. sativa* seed oil on the expression of these markers have been investigated (Cüce et al., 2015). It is understood that *N. sativa* seed oil is effective against diabetes and merits further treatment strategies for preventing apoptosis in vascular structures.

Treatment of streptozotocin-diabetic rats with *N. sativa* extract, *N. sativa* oil, and TQ significantly decreased the diabetes-induced lipid peroxides and hyperglycemia, and significantly increased serum insulin and SOD activity in tissues. *N. sativa* oil and TQ have therapeutic potential and are protective against STZ diabetes by decreasing oxidative stress, thus preserving pancreatic β-cell integrity, leading to increased insulin levels (Abdelmeguid et al., 2010). The protective effects of *N. sativa* oil on insulin sensitivity and ultrastructural changes of pancreatic β-cells in STZ induced diabetic rats are reported (Kanter et al., 2009). It is evident that *N. sativa* treatment exerts a protective effect on diabetes by decreasing morphological changes and preserving the pancreatic β-cell integrity

(Kanter et al., 2009). The anti-hyperglycemic potential of TQ and the effect on the activities of key enzymes of carbohydrate metabolism in STZ-NA–induced diabetic rats have been evaluated (Pari and Sankaranarayanan, 2009). Oral administration of TQ (20, 40, and 80 mg kg^{-1} body weight for 45 days) dose-dependently improved the glycemic status in STZ/NA-induced diabetic rats. The levels of insulin and hemoglobin increased along with a decrease in glucose and Hemoglobin A1c (HbA1c) levels. The altered activities of carbohydrate-metabolizing enzymes were also restored (Pari and Sankaranarayanan, 2009).

In a clinical study, the adjuvant effect of *N. sativa* oil on various clinical and biochemical parameters of the insulin resistance syndrome in patients with diabetes and dyslipidemia have been evidenced (Najmi et al., 2008). *N. sativa* accentuates glucose induced secretion of insulin besides negatively affecting glucose absorption. Hence, it is of immense therapeutic benefit in diabetic individuals (Kapoor, 2009). The effect of *N. sativa* seeds used as an adjuvant therapy in addition to the anti-diabetic medications on the glycemic control of patients with type 2 diabetes mellitus was investigated (Bamosa et al., 2010).

6.4. Insulin Secretion

Thymoquinone and other antioxidant components in *N. sativa* can increase insulin secretion by improving the energy metabolism of mitochondria, and might also reduce liver injury, according to a study conducted on diabetic rats. The compounds also cause an increase in the insulin concentration by promoting the intracellular insulin receptor pathways (Heshmati and Namazi, 2015). It has been suggested that thymoquinone and the other antioxidant components in *N. sativa* can activate the mitogen- activated protein kinases (MAPKs) and protein kinase B (PKB) pathways, which function in insulin sensitivity (Le et al., 2004).

6.5. Gluconeogenesis

N. sativa decreases gluconeogenesis, which contributes to hyperglycemia in diabetic individuals, and thymoquinone can reduce the expression of gluconeogenic enzymes (glucose 6-phosphatase and fructose 1,6-biphosphatase) and the production of hepatic glucose (Heshmati and Namazi, 2015). Furthermore, *N. sativa* prevents gluconeogenesis by activating the protein kinases activated by adenosine monophosphate (AMPK) in liver and muscles (Heshmati et al., 2015).

6.6. Glucose Absorption

It has been shown that liquid intake of *N. sativa* extract reduced glucose absorption and inhibited the glucose carrier in diabetic rats. Another potential mechanism of action is that polyphenol components can suppress the properties of glucose absorption transport (Heshmati and Namazi, 2015). Furthermore, thymoquinone is reported to have a curative effect in decreasing the oxidative stress that results from hyperglycaemia and protects β- cell integrity. As a consequence, the clinical use of thymoquinone can be effective in protecting β-cells against oxidative stress (Kaatabi et al., 2015).

A double-blind placebo controlled study with 114 participants consisting of 63 men and 51 women, studied the effect of *N. sativa* supplement on glycaemic control and antioxidant capacity in Type 2 diabetic patients that used hypoglycaemic drugs; the intervention group was given 2 g/day *N. sativa* for one year. The study revealed considerable decreases in fasting blood glucose and HbA1c levels of individuals; there were significant differences between groups and significant increases were observed in total antioxidant capacity, superoxide dismutase and glutathione levels of the group that took *Nigella sativa* powder compared to the control group. At the end of the treatment, insulin resistance decreased and β-cell activity increased compared to the level in individuals at the start of treatment (Kaatabi et al., 2015).

In a study in which individuals with Type 2 DM were given *N. sativa* powder at a dose of 1, 2 and 3 g/day, a significant decreases were observed in insulin resistance and β-cell function indicators such as postprandial blood sugar, HbA1c, HOMA-IR, only in individuals who were supplied with 2 g *N. sativa* powder per day. The three different doses had no negative effect on the renal or hepatic functions of individuals during the study. The most effective does was determined to be 2 g and this might also confer positive benefits with hypoglycaemic agents (Bamosa et al., 2010). In another study, type 2 DM patients were given 2 g/day *N. sativa* powder for one year and it was shown that HbA1c values decreased significantly in the intervention group (Bamosa, 2015). In another study that treated 60 individuals with insulin resistance who took hypoglycaemic drugs with 2.5 mL *N. sativa* oil twice daily for 60 weeks, significant improvements were observed in TC, LDL and prepandial blood sugar levels. In addition, *N. sativa* oil is effective as an additional treatment in individuals with insulin resistance. *N. sativa* is reported to have important effects especially in diabetic and dyslipidemic individuals (Najmi et al., 2008) and also it has been shown to play an important role in the prevention of diabetic neuropathy in diabetic rats that were supplied with 50 mg/day thymoquinone for 8 weeks (Omran, 2014).

6.7. Anti-Obesity Effects of *N. Sativa*

Obesity is one of the most common health problems in all age groups worldwide. In recent years, herbal supplement use has assumed a place among the complementary diet-based and alternative treatment methods that are commonly used for weight loss (Hasani-Ranjbar et al., 2013). *N. sativa* and its active substance thymoquinone show anti-obesity effects, due to their positive effects against cardiovascular disease, cancer, insulin sensitivity and their immune-modular effects (Vanamala et al., 2012; Ahmad et al., 2013). Although weight loss can occur in individuals, depending on decreases in insulin resistance, *N. sativa* can improve the lipid profile and blood glucose levels in individuals with DM after weight loss (Bamosa et al., 2010). A 25% decrease in food intake was observed in rats that were fed with *N. sativa* oil by intragastric gavage for four weeks. Therefore, *N. sativa* oil might possess anorectic effects, and can cause a decrease in food intake and body weight, and possibly also improvements in lipid peroxidation and insulin sensitivity (Le et al., 2004). Another study demonstrated a significant decreases in the body weight of diabetic rats following treatment with 300 mg/day *N. sativa* extract for 30 days (Fararh et al., 2010). Human studies in which the effect of *N. sativa* on obesity has been examined are limited. In a randomized controlled double-blind single study, 50 obese men were given 3 g/day *Nigella sativa* powder for 3 months and a significant decreases in body weight, and the waist and hip circumference of individuals were found (Datau et al., 2010). However, *N. sativa* given to individuals with normal weight did not cause any significant change in the body weight (Qidwai et al., 2009). It is claimed that the consumption of *N. sativa* might be effective against obesity if used for a long period and in large amounts (Nader et al., 2010). Despite these results, more controlled intervention studies are required to understand better the effects of *N. sativa* on weight loss.

6.8. Anti-Inflammatory Effect of *N. Sativa*

6.8.1. Psoriasis

Psoriasis is common skin condition, which is a hyperproliferative, autoimmune skin disorder and can be itchy and painful. An experimental study was undertaken to see the effect of ethanol extract of *N. sativa* seeds in treatment of psoriasis. It was found that *N. sativa* increases the epidermal thickness when case study group is compared to control group that used traditional treatment (Dwarampudi et al., 2012). Another study was made by Ahmed et al. (2014) to compare asiaticoside, and the ethanolic extract of *N. sativa* was applied in two dosage forms, as an ointment and oral dosage form. They had IC50 value of 23.9 $\mu g\ ml^{-1}$, which is about the IC50 value for asiaticoside (20.13 $\mu g\ ml^{-1}$).

In conclusion, *N. sativa* oil had better effect as antiproliferative activity than the compared treatment. It is concluded based on many researches that *N. sativa* has antipsoriatic effect with the best effect obtained with the combination of ointment and the oral dosage form.

6.8.2. Skin Disease

N. sativa seeds have been used as an external application for different kinds of skin disease for long time. Bhalani and Shah (2015) decided to test *N. sativa* oil antibiotic effect compared with standard drug amoxicillin. The results of both treatments were the same in bacterial zone inhibition. Therefore, *N. sativa* is a good candidate in the treatment of inflamed skin which can be caused by infection, irritation, rashes, dermatitis, acne, and psoriasis.

CONCLUSION

The seeds of *N. sativa* are used in various traditional systems of medicines and folk medicine all over the world for the treatment and prevention of a variety of diseases. *N. sativa* and its active component thymoquinone have positive effects on health, and their mechanism of action depends on the type of disease. Although the degree of effect and mechanism of action of *N. sativa* on some diseases have been demonstrated by *in vivo* and *in vitro* studies. Historical evidence showed the relation between *N. sativa* and human health care system from decay to modern times, which is due to its antimicrobial, anti-inflammatory, and antifungal activity with application for variety of diseases like bronchitis, cough, asthma, hypertension, paralysis, amenorrhea, anorexia, and rheumatism. The scientific data in this chapter will help the researchers to get updated information about *N. sativa*.

REFERENCES

Aboul-Ela, E. I. (2002). Cytogenetic studies on Nigella sativa seeds extract and thymoquinone on mouse cells infected with schistosomiasis using karyotyping. *Mutation Research – Genetic Toxicology and Environmental Mutagenesis*, *516*, 11–17.

Abukhader, M. (2012). The effect of route of administration in thymoquinone toxicity in male and female rats. *Indian Journal of Pharmaceutical Sciences*, *74*, 195.

Adam, G. O., Rahman, M. M., Lee, S. J., Kim, G. B., Kang, H. S., Kim, J. S. and Kim, S. J. (2016). Hepatoprotective Effects of Nigella sativa seed extract against acetaminophen induced oxidative stress. *Asian Pacific Journal of Tropical Medicine*, *9*, 221–227.

Ahmad, S. & Beg, Z. H. (2013). Hypolipidemic and antioxidant activities of thymoquinone and limonene in atherogenic suspension fed rats. *Food Chemistry*, *138*, 1116-1124.

Ahmad, A., Husain, A., Mujeeb, M., Khan, S. A., Najmi, A. K., Siddique, N. A. & Anwar, F. (2013). A review on therapeutic potential of Nigella sativa: A miracle herb. *Asian Pacific Journal of Tropical Biomedicine*, *3*, 337-352.

Ahmed, J. H., Ibraheem, Y. A. & Al-Hamdi, I. K. (2014). Evaluation of efficacy, safety and antioxidant effect of *Nigella sativa* in patients with psoriasis: a randomized clinical trial. *Journal of Clinical and Experimental Investigations.*, *5*(2), 186–193.

Akash, M., Rehman, K., Rasool, F., Sethi, A., Abrar, M., Irshad, A., Abid, A. & Murtaza, G. (2011). Alternate therapy of type 2 diabetes mellitus (T2DM) with *Nigella* (*Ranunculaceae*). *Journal of Medicinal Plants Research*, *5*, 6885-6889.

Aljabre, S. H., Randhawa, M. A., Akhtar, N., Alakloby, O. M., Alqurashi, A. M. & Aldossary, A. (2005). Antidermatophyte activity of ether extract of *Nigella sativa* and its active principle, thymoquinone. *Journal of Ethnopharmacology*, *101*, 116-119.

Akgul, A. (1989). Antimicrobial activity of black cumin (*Nigella sativa* L.) essential oil. Gazi *Journal of Faculty of Pharmacology*, *6*, 63-68.

Ali, B. H. & Blunden, G. (2003). Pharmacological and toxicological properties of *Nigella sativa. Phytopherapy Research*, *17*, 299–305.

Al-Mahasneh, M. A., Ababneh, H. A. & Rababah, T. (2008). Some engineering and thermal properties of black cumin (*Nigella sativa* L.) seeds. *International Journal of Food Science and Technology*, *43*(6), 1047-1052.

American Cancer Society. (2015). *Cancer Facts and Figures 2014* http://www.cancer.org/research/cancerfactsstatistics/cancerfacts figures2014/ (available June, 2016).

Al-Naqeep, G., Al-Zubairi, A. S., Ismail, M., Amom, Z. H. & Esa, N. M. (2011). Antiatherogenic potential of *Nigella sativa* seeds and oil in diet-induced hypercholesterolemia in rabbits. *Evidence- Based Complementary and Alternative Medicine*, 1-8.

Al-Othman, A. M., Ahmad, F., Al-Orf, S., Al-Murshed, K. S. & Ariff, Z. (2006). Effect of dietry supplementation of *Ellataria cardamun* and *Nigella sativa* on the toxicity of rancid corn oil in rats. *International Journal of Pharmacology*, *2*, 60–65.

Aruoma, O. I., Spencer, J. P. E., Warren, D., Jenner, P., Butler, J. & Halliwell, B. (1997). Characterization of food antioxidants, illustrated using commercial garlic and ginger preparations. *Food Chemistry*, *60*(2), 149-156

Asdadi, A., Harhar, H., Gharby, S., Bouzoubaâ, Z., El Yadini, A., Moutaj, R., El Hadek, M., Chebli, B. & Hassani, L. M. I. (2014). Chemical composition and antifungal activity of Nigella Sativa L. oil seed cultivated in Morocco. *International Journal of Pharmaceutical Science Invention*, *3*, 9–15.

Awad, E. M. (2005). *In vitro* decreases of the fibrinolytic potential of cultured human fibrosarcoma cell line, HT1080, by *Nigella sativa* oil. *Phytomedicine*, *12*, 100-107.

Badary, O. A., Al-Shabanah, O. A., Nagi, M. N., Al-Rikabi, A. C. & Elmazar, M. M. (1999). Inhibition of benzo(a)pyrene-induced forestomach carcinogenesis in mice by thymoquinone. *European Journal of Cancer Prevention*, *8*, 435–440.

Badary, O. A. & Gamal El-Din, A. M. (2001). Inhibitory effects of thymoquinone against 20-methylcholanthrene-induced fibrosarcoma tumorigenesis. *Cancer Detection and Prevention*, *25*, 362-368.

Badary, O., Abdel–Naeem, A., Abdel–Wahab, M. & Hamada, F. (2001). The influence of thymoquinone on doxorubicin–induced hyper-lipidemic nephropathy in rats. *Toxicology*, (143), 219–226.

Badary, O. A., Taha, R. A., Gamal el-Din, A. M. & Abdel-Wahab, M. H. (2003). Thymoquinone is a potent superoxide anion scavenger. *Drug and Chemical Toxicology*, *26*, 87-98.

Bakathir, H. A. & Abbas, N. A. (2011). Detection of the antibacterial effect of *Nigella sativa* ground seeds with water. *African Journal of Traditional, Complementary and Alternative Medicines.*, *8*, 159-164.

Bhalani, U. & Shah, K. (2015). Preparation and evaluation of topical gel of *Nigella sativa* (KALONJI). *International Journal of Research and Development in Pharmacy & Life Sciences*, *4*(4), 1669–1672.

Bamosa, A. O., Ali, B. A. & Al-Hawsawi, Z. A. (2002). The effect of thymoquinone on blood lipids in rats. *Indian Journal of Physiology and Pharmacology*, *46*, 195-201.

Bamosa, A. O., Kaatabi, H., Lebdaa, F. M., Elq, A. M. & Al-Sultanb, A. (2010). Effect of *Nigella sativa* seeds on the glycemic control of patients with type 2 diabetes mellitus. *Indian Journal of Physiology and Pharmacology*, *54*, 344–354.

Bamosa, A. O. (2015). A review on the hypoglycemic effect of *Nigella sativa* and thymoquinone. *Saudi Journal of Medicine and Medical Sciences*, *3*, 2.

Balbaa, M., El-Zeftawy, M., Ghareeb, D., Taha, N. & Mandour, A. W. (2016). *Nigella sativa* relieves the altered insulin receptor signaling in streptozotocin-induced diabetic rats fed with a high-fat diet. *Oxidative Medicine and Cellular Longevity*, 2492107.

Banerjee, S., Kaseb, A. O., Wang, Z., Kong, D., Mohammad, M., Padhye, S., Sarkar, F. H. & Mohammad, R. M. (2009). Antitumor activity of gemcitabine and oxaliplatin is augmented by thymoquinone in pancreatic cancer. *Cancer Research*, *69*, 5575-5583.

Bita, A., Rosu, A., Calina, D., Rosu, L., Zlatian, O., Dindere, C. & Simionescu, A. (2012). An alternative treatment for Candida infections with Nigella sativa extracts. *European Journal of Hospital Pharmacy: Science and Practice*, *19*, 162.

Barakat, E. M. F., El Wakeel, L. M. & Hagag, R. S. (2013). Effects of Nigella sativaon outcome of hepatitis C in Egypt. *World Journal of Gastroenterology*, *19*, 2529–2536.

Benhaddou-Andaloussi, A., Martineau, L. C., Spoor, D., Vuong, T., Leduc, C., Joly, E., Burt, A., Meddah, B., Settaf, A., Arnason, J. T., Prentki, M., Haddad, P. S. & Arnason, J. T. (2008). Antidiabetic activity of *Nigella sativa* seed extract in cultured pancreatic β-cells, skeletal muscle cells, and adipocytes. *Pharmaceutical Biology*, *46*, 96-104.

Bordoni, L, Fedeli, D., Nasuti, C., Maggi, F., Papa, F., Wabitsch, M., De Caterina, R. & Gabbianelli, R. (2019). Antioxidant and Anti-Inflammatory Properties of *Nigella sativa* Oil in Human Pre-Adipocytes. *Antioxidants (Basel)*, *8*(2), 51.

Burits, M. & Bucar, F. (2000). Antioxidant activity of *Nigella sativa* essential oil. *Phytotherapy Research*, *14*, 323–328.

Chaieb, K., Kouidhi, B., Jrah, H., Mahdouani, K. & Bakhrouf, A. (2011). Antibacterial activity of Thymoquinone, an active principle of *Nigella sativa* and its potency to prevent bacterial biofilm formation. *BMC Complementary and Alternative Medicine.*, *11*, 29.

Chehl, N., Chipitsyna, G., Gong, Q., Yeo, C. J. & Arafat, H. A. (2009). Anti-inflammatory effects of the *Nigella sativa* seed extract, thymoquinone, in pancreatic cancer cells. *HPB*, *11*, 373-381.

Cüce, G., Sözen, M. E., Çetinkaya, S., Canbaz, H. T., Seflek, H. & Kalkan, S. (2015). Effects of *Nigella sativa* L. seed oil on intima-media thickness and Bax and Caspase-3 expression in diabetic rat aorta. *Anatolian Journal of Cardiology*, *16*, 460–446.

Darakhshan, S., Pour, A. B., Colagar, A. H. & Sisakhtnezhad, S. (2015). Thymoquinone and its therapeutic potentials. *Pharmacological Research*, *95*, 138-158.

Datau, E., Surachmanto, E., Pandelaki, K. & Langi, J. (2010). Efficacy of Nigella sativa on serum free testosterone and metabolic disturbances in central obese male. *Acta Medica Indonesiana*, *42*, 130-134.

De, M., De, A. K. & Banerjee, A. B. (1999). Antimicrobial screening of some Indian Spices. *Phytotherapy Research*, *13*, 616–618.

Dehkordi, F. R. & Kamkhah, A. F. (2008). Antihypertensive effect of *Nigella sativa* seed extract in patients with mild hypertension. *Fundamental and Clinical Pharmacology*, *22*, 447-452.

Deepak, S. S., Sikender, M., Garg, V. & Samim, M. (2011). Entrapment of seed extract of *Nigella sativa* into thermosensitive (NIPAAm–Co–VP) co-polymeric micelles and its antibacterial activity. *International Journal of Pharmaceutical Sciences and Drug Research*, *3*, 246-252.

Dorman, H. J. & Deans, S. G. (2000). Antimicrobial agents from plants: antibacterial activity of plant volatile oils. *Journal of Applied Microbiology*, *88*, 308-316.

Dwarampudi, L. P., Palaniswamy, D., Nithyanantham, M. & Raghu, P. S. (2012). Antipsoriatic activity and cytotoxicity of ethanolic extract of *Nigella sativa* seeds. *Pharmacognosy Magazine*, *8*(32), 268–272.

Effenberger, K., Breyer, S. & Schobert, R. (2010). Terpene conjugates of the Nigella sativa seed-oil constituent thymoquinone with enhanced efficacy in cancer cells. *Chemistry & Biodiversity*, *7*, 129-139.

El-Mahdy, M. A., Zhu, Q., Wang, Q. E., Wani, G. & Wani, A. A. (2005). Thymoquinone induces apoptosis through activation of caspase-8 and mitochondrial events in p53-null myeloblastic leukemia HL-60 cells. *International Journal of Cancer*, *117*, 409-417.

El-Mahmoudy, A., Matsuyama, H., Borgan, M., Shimizu, Y., El-Sayed, M., Minamoto, N. & Takewaki, T. (2002). Thymoquinone suppresses expression of inducible nitric oxide synthase in rat macrophages. *International Immunopharmacology*, *2*(11), 1603-1611.

El-Nagerabia, S. A., Al-Bahryb, S. N., Elshafieb, A. E. & AlHilalib, S. (2012). Effect of *Hibiscus sabdariffa* extract and *Nigella sativa* oil on the growth and aflatoxin B1 production of *Aspergillus flavus* and *Aspergillus parasiticus* strains. *Food Control*, *25*, 59-63.

El Shenawy, N. S., Soliman, M. F. M. & Reyad, S. I. (2008). The effect of antioxidant properties of aqueous garlic extract and *Nigella sativa* as anti-schistosomiasis agents in mice. *Revista do Instituto de Medicina Tropical de S˜ao Paulo*, *50*, 29–36.

Entok, E., Ustuner, M. C., Ozbayer, C., Tekin, N., Akyuz, F., Yangi, B., Kurt, H., Degirmenci, I. & Gunes, H. V. (2014). Anti-inflammatuar and anti-oxidative effects of *Nigella sativa* L.: 18FDG-PET imaging of inflammation. *Molecular Biology Reports*, *41*, 2827-2834.

El-Dakhakhny, M., Mady, N., Lembert, N. & Ammon, H. P. (2002). The hypoglycemic effect of *Nigella sativa* oil is mediated by extrapancreatic actions. *Planta Medica*, *68*, 465–466.

Emeka, L. B., Emeka, P. M. & Khan, T. M. (2015). Antimicrobial activity of *Nigella sativa* L. seed oil against multi-drug resistant *Staphylococcus aureus* isolated from diabetic wounds. *Pakistan Journal of Pharmaceutical Sciences*, *28*, 1985–1990.

Fallah Huseini, H., Amini, M., Mohtashami, R., Ghamarchehre, M., Sadeqhi, Z., Kianbakht, S. & Fallah Huseini, A. (2013). Blood pressure lowering effect of *Nigella sativa* L. Seed oil in healthy volunteers: A Randomized, double-blind, placebo controlled clinical trial. *Phytotherapy Research*, *27*, 1849-1853.

Farah, I. O. & Begum, R. A. (2002). Effect of Nigella sativa (*N. sativa* L.) and oxidative stress on the survival pattern of MCF-7 breast cancer cells. *Biomedical Sciences Instrumentation*, *39*, 359-364.

Fararh, K. M., Atoji, Y., Shimizu, Y. & Takewaki, T. (2002). Isulinotropic properties of *Nigella sativa* oil in streptozotocin plus nicotinamide diabetic hamster. *Research in Veterinary Science*, *73*, 279–282.

Fararh, K. M., Atoji, Y., Shimizu, Y., Shiina, T., Nikami, H. & Takewaki, T. (2004). Mechanisms of the hypoglycaemic and immunopotentiating effects of *Nigella sativa* L. Oil in streptozotocin-induced diabetic hamsters. *Research in Veterinary Science*, *77*, 123–129.

Fararh, K. M., Ibrahim, A. K. & Elsonosy, Y. A. (2010). Thymoquinone enhances the activities of enzymes related to energy metabolism in peripheral leukocytes of diabetic rats. *Research in Veterinary Science*, *88*, 400-404.

Farrag, H. A., El-Bazza, Z. E. M., El-Fouly, M. E. D. & El-Tablawy, S. Y. M. (2000). Effect of gamma radiation on the bacterial flora of *Nigella sativa* seeds and its oil constituents. *Acta Pharma*, *50*, 195-207.

Forouzanfar, F., Bazzaz, B. S. F. & Hosseinzadeh, H. (2014). Black cumin (*Nigella sativa*) and its constituent (thymoquinone): A review on antimicrobial effects. *Iranian Journal of Basic Medical Sciences*, *17*, 929–938.

Gali-Muhtasib, H., Diab-Assaf, M., Boltze, C., Al-Hmaira, J., Hartig, R., Roessner, A. & Schneider-Stock, R. (2004). Thymoquinone extracted from black seed triggers apoptotic cell death in human colorectal cancer cells via a p53-dependent mechanism. *International Journal of Oncology*, *25*, 857–866.

Gendy, E., Hessien, M., Abdel Salamm, I., Moradm, M. E. L., Magrabym, K., Ibrahimm, H. A., Kalifa, M. H. & El-Aaser, A. A. (2007). Evaluation of the possible antioxidant effects of Soybean and *Nigella sativa* during experimental hepatocarcinogenesis by nitrosamine precursors. *Turkish Journal of Biochemistry*, *32*, 5–11.

Güllü, E. B. & Gülcan, A. (2013). Timokinon: *Nigella sativa*' nın Biyoaktif Komponenti. *Kocatepe Veterinary Journal*, *6*, 51-61.

Gupta, S., Satishkumar, M., Duraiswamy, B., Das, S. & Chhajed, M. (2012). Potential herbs and its phytoconstituents against fungal infection: a systematic review. *World Journal of Pharmaceutical Research*, *1*, 1–20.

Halawani, E. (2009). Antibacterial activity of thymo-quinone and thymohydroquinone of *Nigella sativa* L. and their interaction with some antibiotics. *Advanced Biomedical Research*, *3*, 148-152.

Hanafy, M. S. M. & Hatem, M. E. (1991). Studies on the antimicrobial activity of (black cumin). *Journal of Ethnopharmacology*, *34*, 275-278.

Hannan, A., Saleem, S., Chaudhary, S., Barkaat, M. & Arshad, M. U. (2008). Anti-bacterial activity of *Nigella sativa* against clinical isolates of methicillin resistant *Staphylococcus aureus*. *Journal of Ayub Medical College, Abbottabad*, *20*, 72-74.

Harzallah, H. J., Kouidhi, B., Flamini, G., Bakhrouf, A. & Mahjoub, T. (2011). Chemical composition, antimicrobial potential against cariogenic bacteria and cytotoxic activity of Tunisian *Nigella sativa* essential oil and thymoquinone. *Food Chemistry*, *129*, 1469–1474.

Harzallah, H. J., Grayaa, R., Kharoubi, W., Maaloul, A., Hammami, M. & Mahjoub, T. (2012). Thymoquinone, the *Nigella sativa* bioactive compound, prevents circulatory oxidative stress caused by 1,2-dimethylhydrazinein erythrocyte during colon postinitiation carcinogenesis. *Oxidative Medicine and Cellular Longevity*, 854065.

Harzallah, H. J., Noumi, E., Bekir, K. Bakhrouf, A. & Mahjoub, T. (2012). Chemical composition, antibacterial and antifungal properties of Tunisian Nigella sativa fixed oil. *African Journal of Microbiology Research.*, *6*, 4675–4679.

Hasan, N. A., Nawahwi, M. Z. & Malek, H. A. (2013). Antimicrobial activity of *Nigella sativa* seed extract. *Sains Malaysiana*, *42*, 143-147.

Hasani-Ranjbar, S., Jouyandeh, Z. & Abdollahi, M. (2013). A systematic review of antiobesity medicinal plants-an update. *Journal of Diabetes and Metabolic Disorders*, *12*, 28.

Heshmati, J. & Namazi, N. (2015). Effects of black seed (*Nigella sativa*) on metabolic parameters in diabetes mellitus: A systematic review. *Complementary Therapies in Medicine*, *23*, 275-282.

Heshmati, J., Namazi, N., Memarzadeh, M. R., Taghizadeh, M. & Kolahdooz, F. (2015). *Nigella sativa* oil affects glucose metabolism and lipid concentrations in patients with type 2 diabetes: A randomized, double-blind, placebo controlled trial. *Food Research International*, *70*, 87-93.

Hosseinzadeh, H. & Parvardeh, S. (2004). Anticonvulsant effects of thymoquinone, the major constituent of *Nigella sativa* seeds, in mice. *Phytomedicine*, *11*, 56-64.

Hosseinzadeh, H., Parvardeh, S., Asl, M. N., Sadeghnia, H. R. & Ziaee, T. (2007). Effect of thymoquinone and *Nigella sativa* seeds oil on lipid peroxidation level during global cerebral ischemia-reperfusion injury in rat hippocampus. *Phytomedicine: International Journal of Phytotherapy and Phytopharmacology*, *14*, 621–627.

Javed, S., Shahid, A. A., Haider, M. S., Umeera, A., Ahmad, R. & Mushtaq, S. (2012). Nutritional, phytochemical potential and pharmacological evaluation of *Nigella Sativa* (Kalonji) and *Trachyspermum Ammi* (Ajwain). *Journal of Medicinal Plants Research*, *6*, 768–775.

Islam, M. H., Ahmad, I. Z. & Salman, M. T. (2013). Antibacterial activity of Nigella sativa seed in various germination phases on clinical bacterial strains isolated from human patients. *E3 Journal of Biotechnology and Pharmaceutical Research*, *4*, 8–13.

Kaatabi, H., Bamosa, A. O., Lebda, F. M., Al Elq, A. H. & Al-Sultan, A. I. (2012). Favorable impact of Nigella sativa seeds on lipid profile in type 2 diabetic patients. *Journal of Family and Community Medicine*, *19*, 155.

Kaatabi, H., Bamosa, A. O., Badar, A., Al-Elq, A., Abou-Hozaifa, B., Lebda, F., Al-Khadra, A. & Al-Almaie, S. (2015). *Nigella sativa* improves glycemic control and ameliorates oxidative stress in patients with type 2 diabetes mellitus: Placebo controlled participant blinded clinical trial. *PloS one*, *10*, e0113486.

Kaleem, M., Kirmani, D., Asif, M., Ahmed, Q. & Bano, B. (2006). Biochemical effects of *Nigella sativa* L. seeds in diabetic rats. *Indian Journal of Experimental Biology*, *44*, 745–748.

Kanter, M., Meral, I., Dede, S., Gunduz, H., Cemek, M., Ozbek, H. & Uygan, I. (2003). Effects of *Nigella sativa* L. and urtica dioica L. on lipid peroxidation, antioxidant enzyme systems and some liver enzymes in ccl4-treated rats. *Journal of Veterinary Medicine. A, Physiology, Pathology, Clinical Medicine*, *50*, 264–268.

Kanter, M., Coskun, O., Korkmaz, A. & Oter, S. (2004). Effects of *Nigella sativa* on oxidative stress and beta-cell damage in streptozotocin-induced diabetic rats. *The Anatomical Record. Part A, Discoveries in Molecular, Cellular, and Evolutionary Biology*, *279*, 685–691.

Kanter, M. (2009). Effects of *Nigella sativa* seed extract on ameliorating lung tissue damage in rats after experimental pulmonary aspirations. *Acta Histochemica*, *111*, 393–403.

Kapoor, S. (2009). Emerging clinical and therapeutic applications of nigella sativa in gastroenterology. *World Journal of Gastroenterology*, *15*, 2170–2171.

Kaseb, A. O., Chinnakannu, K., Chen, D., Sivanandam, A., Tejwani, S., Menon, M., Dou, Q. P. & Reddy, G. P. (2007). Androgen receptor and E2F-1 targeted thymoquinone therapy for hormone-refractory prostate cancer. *Cancer Research*, *67*, 7782-7788.

Kaya, M. S., Kara, M. & Özbek, H. (2003). The effects of black cumin (*Nigella sativa*) seeds on CD^{3+}, CD^{4+}, CD^{8+} cells of human immune system and total number of leucocytes. *Genel Tıp Derg*, *13*, 109-112.

Keyhanmanesh, R., Gholamnezhad, Z. & Boskabady, M. H. (2014). The relaxant effect of Nigella sativa on smooth muscles, its possible mechanisms and clinical applications. *Iranian Journal of Basic Medical Sciences*, *17*, 939.

Khader, M., Bresgen, N. & Eckl, P. (2010). Antimutagenic effects of ethanolic extracts from selected Palestinian medicinal plants. *Journal of Ethnopharmacology*, *127*, 319-324.

Khan, M. A. U., Ashfaq, M. K., Zuberi, H. S., Mahmood, M. S. & Gilani, A. H. (2003). The *in vivo* antifungal activity of the aqueous extract from *Nigella sativa* seeds. *Phytotherapy Research*, *17*, 183–186.

Khan, N. & Sultana, S. (2005). Inhibition of two stage renal carcinogenesis, oxidative damage and hyperproliferative response by *Nigella sativa. European Journal of Cancer Prevention: the Official Journal of the European Cancer Prevention Organisation (ecp)*, *14*, 159–168.

Khan, M. A., Chen, H. C., Tania, M. & Zhang, D. Z. (2011). Anticancer activities of Nigella sativa (black cumin). *African Journal of Traditional, Complementary and Alternative medicines*, *8*, 226–232.

Khan, M. A., Tania, M., Fu, S. & Fu, J. (2017). Thymoquinone, as an anticancer molecule: from basic research to clinical investigation. *Oncotarget*, *8*(31), 51907-51919.

Khosravi, A. R., Minooeianhaghighi, M. H., Shokri, H., Emami, S. A., Alavi, S. M. & Asili, J. (2011). The potential inhibitory effect of *Cuminum cyminum*, *Ziziphora clinopodioides* and *Nigella sativa* essential oils on the growth of *Aspergillus fumigatus* and *Aspergillus flavus*. *Brazilian Journal of Microbiology*, *42*, 216–224.

Kocyigit, Y., Atamer, Y. & Uysal, E. (2009). The effect of dietary supplementation of *Nigella sativa* L. on serum lipid profile in rats. *Saudi Medical Journal*, *30*, 893-896.

Ibrahim, R. M., Hamdan, N. S., Ismail, M., Saini, S. M., Rashid, S. A., Latiff, L. A. & Mahmud, R. (2014). Protective effects of *Nigella sativa* on metabolic syndrome in menopausal women. *Advanced Pharmaceutical Bulletin*, *4*, 29-33.

Latiff, L. A., Parhizkar, S., Dollah, M. A. & Hassan, S. T. S. (2014). Alternative supplement for enhancement of reproductive health and metabolic profile among perimenopausal women: a novel role of Nigella sativa. *Iranian Journal of Basic Medical Sciences*, *17*, 980.

Le, P., Benhaddou-Andaloussi, A., Elimadi, A., Settaf, A., Cherrah, Y. & Haddad, P. (2004). The petroleum ether extracts of *Nigella sativa* seeds exert insulin sensitizing and lipid lowering action in rats. *Journal of Ethnopharmacology*, *94*, 251-259.

Leong, X. F., Rais Mustafa, M. & Jaarin, K. (2013). *Nigella sativa* and its protective role in oxidative stress and hypertension. *Evidence-Based Complementary and Alternative Medicin*, *201*, 1-9.

Mabrouk, G. M., Moselhy, S. S., Zohny, S. F., Ali, E. M., Helal, T. E., Amin, A. A. & Khalifa, A. A. (2002). Inhibition of methylnitrosourea (MNU) induced oxidative stress and carcinogenesis by orally administered bee honey and *Nigella* grains in Sprague Dawely rats. *Journal of Experimental & Clinical Cancer Research*, *21*, 341-346.

Mahmoud, M. R., El-Abhar, H. S. & Saleh, S. (2002). The effectof *Nigella sativa* oil against the liver damage induced by *Schistosoma mansoni* infection in mice," *Journal of Ethnopharmacology*, *79*, 1–11.

Mahmoudvand, H., Sepahvand, A., Jahanbakhsh, S., Ezatpour, B. & Mousavi, S. A. A. (2014). Evaluation of antifungal activities of the essential oil and various extracts of *Nigella sativa* and its main component, thymoquinone against pathogenic dermatophyte strains. *Journal of Medical Mycology*, *24*, e155–e161.

Mani, P. & Rohatgi, A. (2015). Niacin therapy, HDL cholesterol, and cardiovascular disease: is the HDL hypothesis defunct? *Current Atherosclerosis Reports*, *17*, 1-9.

Mbarek, A. L., Mouse, A. H., Elabbadi, N., Bensalah, M., Gamouh, A., Aboufatima, R., Benharref, A., Chait, A., Kamal, M., Dalal, A. & Zyad, A. (2007). Anti-tumor properties of blackseed (*Nigella sativa* L.) extracts. *Brazilian Journal of Medical and Biological Research*, *40*, 839-847.

Meral, I., Yener, Z., Kahraman, T. & Mert, N. (2001). Effect of *Nigella sativa* on glucose concentration, lipid peroxidation, anti-oxidant defence system and liver damage in experimentally-induced diabetic rabbits. *Journal of Veterinary Medicine. A, Physiology, Pathology, Clinical Medicine*, *48*, 593–599.

Mohamed, A. M., Metwally, N. M. & Mahmoud, S. S. (2005). Sativa seeds against Schistosoma mansoni different stages. *Memorias do Instituto Oswaldo Cruz*, *100*, 205–211.

Monika, T., Sasikala, P., Bhaskara, V. & Reddy, M. (2013). A investigational of antibacterial activities of *Nigella sativa* on mastaitis in dairy crossbred cows. *International Journal of Advanced Scientific and technical Research*, *3*, 263-272.

Morsi, N. M. (2000). Antimicrobial effect of crude extracts of *Nigella sativa* on multiple antibiotics-resistant bacteria. *Acta Microbiologica Polonica*, *49*, 63–74.

Nader, M. A., El-Agamy, D. S. & Suddek, G. M. (2010). Protective effects of propolis and thymoquinone on development of atherosclerosis in cholesterol-fed rabbits. *Archives of Pharmacal Research (Seoul)*, *33*, 637.

Nagi, M. N. & Mansour, M. A. (2000). Protective effect of thymoquinone against doxorubicin-induced cardiotoxicity in rats: a possible mechanism of protection. *Pharmacology Research*, *41*, 283–289.

Nagi, M. N. & Almakki, H. A. (2009). Thymoquinone supplementation induces quinone reductase and glutathione transferase in mice liver: possible role in protection against chemical carcinogenesis and toxicity. *Phytotherapy Research*, *23*, 1295-1298.

Nair, M. K. M., Vasudevan, P. & Venkitanarayanan, K. (2005). Antibacterial effect of black seed oil on *Listeria monocytogenes*. *Food Control*, *16*, 395–398.

Najmi, A., Haque, S. F., Naseeruddin, M. & Khan, R. A. (2008). Effect of *Nigella sativa* oil on various clinical and biochemical parameters of metabolic syndrome. *International Journal of Diabetes in Developing Countries*, *16*, 85–87.

Nokoff, N. J., Rewers, M. & Cree Green, M. (2012). The interplay of autoimmunity and insulin resistance in type 1 diabetes. *Discovery Medicine*, *13*, 115–22.

Norwood, A. A., Tan, M., May, M., Tucci, M. & Benghuzzi, H. (2006). Comparison of potential chemotherapeutic agents, 5-fluoruracil, green tea, and thymoquinone on colon cancer cells. *Biomedical sciences instrumentation*, *42*, 350-356.

Omran, O. M. (2014). Effects of thymoquinone on STZ-induced diabetic nephropathy: an immunohistochemical study. *Ultrastructural Pathology*, *38*, 26-33.

Paarakh, P. M. (2010). *Nigella sativa* Linn. *A Comprehensive Review.*, *1*, 409-429.

Pari, L. & Sankaranarayanan, C. (2009). Beneficial effects of thymoquinone on hepatic key enzymes in streptozotocin-nicotinamide induced diabetic rats. *Life Sciences*, *85*, 830–834.

Peterhans, E. (1997). Oxidants and antioxidants in viral diseases: disease mechanisms and metabolic regulation. *Journal of Nutrition*, *127*, 962S– 965S.

Piras, A., Rosa, A., Marongiu, B. Porceddaa, S., Falconieric, D., Dessìb, M. A., Ozcelikd, B. & Koca, U. (2013). Chemical composition and *in vitro* bioactivity of the volatile and fixed oils of *Nigella sativa* L. extracted by supercritical carbon dioxide. *Industrial Crops and Products*, *46*, 317–323.

Qidwai, W., Hamza, H. B., Qureshi, R. & Gilani, A. (2009). Effectiveness, safety, and tolerability of powdered *Nigella sativa* (kalonji) seed in capsules on serum lipid levels, blood sugar, blood pressure, and body weight in adults: results of a randomized, double-blind controlled trial. *The Journal of Alternative and Complementary Medicine*, *15*, 639-644.

Ragheb, A., Elbarbry, F., Prasad, K., Mohamed, A., Ahmed, M. S. & Shoker, A. (2008). Attenuation of the development of hypercholesterolemic atherosclerosis by thymoquinone. *International Journal of Angiology*, *17*, 186-192.

Ramadan, M. F. (2007). Nutritional value, functional properties and nutraceuticalb applications of black cumin (*Nigella sativa* L.): an overview. *International Journal of Food Science and Technology*, *42*, 1208–1218.

Randhawa, M. A. & Alghamdi, M. S. (2011). Anticancer activity of *Nigella sativa* (black seed)—a review. *The American Journal of Chinese Medicine*, *39*, 1075-1091.

Razavi, B. & Hosseinzadeh, H. (2014). A review of the effects of *Nigella sativa* L. and its constituent, thymoquinone, in metabolic syndrome. *Journal of Endocrinological Investigation*, *37*, 1031-1040.

Rchid, H., Chevassus, H., Nmila, R., Guiral, C., Petit, P., Chokaïri, M. & Sauvaire, Y. (2004). *Nigella sativa* seed extracts enhance glucose-induced insulin release from rat-isolated langerhans islets. *Fundamental and Clinical Pharmacology*, *18*, 525–529.

Rogozhin, E. A., Oshchepkova, Y. I., Odintsova, T. I., Khadeeva, N. V., Veshkurova, O. N., Egorov, T. A., Grishin, E. V. & Salikhov, S. I. (2011). Novel antifungal defensins from *Nigella sativa* L. seeds. *Plant Physiol Biochem*, *49*, 131-137.

Rooney, S. & Ryan, M. (2005). Modes of action of alpha-hederin and thymoquinone, active constituents of *Nigella sativa*, against HEp-2 cancer cells. *Anticancer Research*, *25*, 4255-4259.

Sagdic, O., Kuscu, A., Ozcan, M. & Ozcelik, S. (2002). Effects of Turkish spice extracts at various concentrations on the growth of *Escherichia coli* O157:H7. *Food Microbiology*, *19*, 473-480.

Sagdic, O. (2003). Sensitivity of four pathogenic bacteria to Turkish thyme and oregano hydrosols. *LWT - Food Scienc and Technology*, *36*, 467-473.

Şakalar, Ç., İzgi, K., İskender, B., Sezen, S., Aksu, H., Çakır, M., Kurt, B., Turan, A. & Canatan, H. (2016). The combination of thymoquinone and paclitaxel shows anti-tumor activity through the interplay with apoptosis network in triplenegative breast cancer. *Tumour Biology*, *37*, 4467–4477.

Salim, E. I. & Fukushima, S. (2003). Chemopreventive potential of volatile oil from black cumin (*Nigella sativa* L.) seeds against rat colon carcinogenesis. *Nutrition and Cancer*, *45*, 195-202.

Salem, M. L. & Hossain, M. S. (2000). *In vivo* acute depletion of CD8 (+) T cells before murine cytomegalovirus infection upregulated innate antiviral activity of natural killer cells. *International Journal of Immunopharmacology*, *22*, 707–718.

Salem, M. L. (2005). Immunomodulatory and therapeutic properties of the *Nigella sativa* L. seed. *International immunopharmacology*, *5*, 1749-1770.

Salem, E. M., Yar, T., Bamosa, A. O., Al-Quorain, A., Yasawy, M. I., Alsulaiman, R. M. & Randhawa, M. A. (2010). Comparative study of Nigella Sativa and triple therapy in eradication of *Helicobacter Pylori* in patients with non-ulcer dyspepsia. *Saudi Journal of Gastroenterology*, *16*, 207–214.

Salem, A. A., El Haty, I. A., Abdou I. M. & Mu, Y. (2015). Interaction of human telomeric G-quadruplex DNA with thymoquinone: a possible mechanism for thymoquinone anticancer effect. *Biochemica and Biophysica Acta*, *1850*, 329–342

Salama, R. H. (2010). Clinical and therapeutic trials of *Nigella sativa*. *TAF Preventive Medicine Bulletin*, *9*, 513-522.

Salomi, M. J., Nair, S. C. & Panikkar, K. R. (1991). Inhibitory effects of *Nigella sativa* and saffron (*Crocus sativus*) on chemical carcinogenesis in mice. *Nutrition and Cancer*, *16*, 67-72.

Shafi, G., Munshi, A., Hasan, T. N., Alshatwi, A. A., Jyothy, A. & Lei, D. K. (2009). Induction of apoptosis in HeLa cells by chloroform fraction of seed extracts of *Nigella sativa*. *Cancer Cell International*, *9*, 29-36.

Shafiq, H., Ahmad, A., Masud, T. & Kaleem, M. (2014). Cardio-protective and anti-cancer therapeutic potential of *Nigella sativa*. *Iranian Journal of Basic Medical Sciences*, *17*, 967.

Shoieb, A. M., Elgayyar, M., Dudrick, P. S., Bell, J. L. & Tithof, P. K. (2003). *In vitro* inhibition of growth and induction of apoptosis in cancer cell lines by thymoquinone. *International Journal of Oncology*, *22*, 107-114.

Singh, S., Das, S., Singh, G., Schuff, C., de Lampasona, M. P. & Catalán, C. A. (2014). Composition, *in vitro* antioxidant and antimicrobial activities of essential oil and oleoresins obtained from black cumin seeds (*Nigella sativa* L.). *BioMed research international*, 918209.

Sogut, B., Celik, I. & Tuluce, Y. (2008). The effects of diet supplemented with the black Cumin (*Nigella sativa* L.) upon immune potential and antioxidant marker enzymes and lipid peroxidation in broiler chicks. *Journal of Animal and Veterinary Advances*, *7*, 1196–1199.

de Souza Grinevicius, V. M., Kviecinski, M. R., Santos Mota, N. S., Ourique, F., Porfirio Will Castro, L. S., Andreguetti, R. R., et al., (2016). Piper nigrum ethanolic extract rich in piperamides causes ROS overproduction, oxidative damage in DNA leading to cell cycle arrest and apoptosis in cancer cells. *Journal of Ethnopharmacology*, *189*, 139-47.

Sultan, M. T., Butt, M. S., Anjum, F. M., Jamil, A., Akhtar, S. & Nasir, M. (2009). Nutritional profile of indigenous cultivar of black cumin seeds and antioxidant potential of its fixed and essential oil. *Pakistan Journal of Botany*, *41*, 1321-1330.

Sunita, M. & Meenakshi, S. H. (2013). Chemical composition and antidermatophytic activity of *Nigella sativa* essential oil. *African Journal of Pharmacy and Pharmacology*, *7*, 1286-1292.

Suthar, M., Patel, P., Shah, T. & Patel, R. K. (2010). *In vitro* screening of *Nigella sativa* seeds for antifungal activity. *International Research Journal of Pharmaceutical and Applied Sciences*, vol. *1*, 84–91.

Swamy, S. M. & Huat, B. T. (2003). Intracellular glutathione depletion and reactive oxygen species generation are important in alphahederin-induced apoptosis of P388 cells. *Molecular and Cellular Biochemistry*, *24*, 127-139.

Takruri, H. R. H. & Dameh, M. A. F. (1998). Study of the nutritional value of black cumin seeds (*Nigella sativa* L.). *Journal of the Science of Food and Agriculture*, *76*, 404-410.

Tembhurne, S., Feroz, S., More, B. & Sakarkar, D. (2014). A review on therapeutic potential of *Nigella sativa* (kalonji) seeds *Journal of Medicinal Plants Research*, *8*, 167–177.

Thabrew, M. I., Mitry, R. R., Morsy, M. A. & Hughes, R. D. (2005). Cytotoxic effects of a decoction of *Nigella sativa*, Hemidesmus indicus and Smilax glabra on human hepatoma HepG2 cells. *Life Science*, *77*, 1319-1330.

Tingfang, Y., Sung-Gook, C., Zhengfang, Yi., Xiufeng, Pang., Melissa R., Ying, W., Gautam, S., Bharat, B. A. & Mingyao, L. (2008). Thymoquinone inhibits tumor angiogenesis and tumor growth through suppressing AKT and extracellular signal-regulated kinase signaling pathways. *Molecular Cancer Therapeutics*, *7*(7), 1789–96.

Torres, M. P., Ponnusamy, M. P., Chakraborty, S., Smith, L. M., Das, S., Arafat, H. A. & Batra, S. K. (2010). Effects of Thymoquinone in the Expression of Mucin 4 in Pancreatic Cancer Cells: Implications for the development of novel Cancer Therapies. *Molecular Cancer Therapeutics*, *9*, 1419-1431.

Tüfek, N. H., Altunkaynak, M. E., Altunkaynak, B. Z. & Kaplan, S. (2015). Effects of thymoquinone on testicular structure and sperm production in male obese rats. *Systems Biology in reproductive Medicine*, *61*, 194-204.

Türkoğlu, S., Bulmuş, F. G., Parmaksiz, A., Özkan, Y. & Gürsu, F. (2008). Metabolik sendromlu hastalarda paraoksonaz 1 ve arilesteraz aktivite düzeyleri. *Fırat Tıp Dergisi*, *13*, 110-115.

Ustum, G., Kent, L., Chekin, N. & Civelekoglu, H. (1990). Investigation of the technological properties of *Nigella sativa* (black cumin) seeds oil. *Association of Official Analytical Chemists*, *67*, 958-960.

Vanamala, J., Kester, A. C., Heuberger, A. L. & Reddivari, L. (2012). Mitigation of obesitypromoted diseases by Nigella sativa and thymoquinone. *Plant Foods for Human Nutrition*, *67*, 111-119.

WHO. (2015). *Cancer*. http://www.who.int/mediacentre/factsheets /fs297/en/ (available June, 2016).

Woo, C. C., Kumar, A. P., Sethi, G. & Ten, K. H. B. (2012). Thymoquinone: potential cure for inflammatory disorders and cancer. *Biochemcal Pharmacology*, *83*, 443–451.

Yakup, K. (2007). Samsun yöresinde ve misir ülkesinde yetiştirilen çörekotu (*Nigella sativa* L.) Tohumlarinin antioksidan aktivite yönünden incelenmesi. *Süleyman demirel Üniversitesi Fen Dergisi*, *2*, 197-203. [Investigation of nigella (Nigella sativa L.) seeds grown in Samsun region and corn country in terms of antioxidant activity. *Süleyman demirel University Journal of Science*]

Yi, T., Cho, S. G., Yi, Z., Pang, X., Rodriguez, M., Wang, Y., Sethi, G., Aggarwal, B. B. & Liu, M. (2008). Thymoquinone inhibits tumor angiogenesis and tumor growth through suppressing AKT and extracellular signal-regulated kinase signaling pathways. *Molecular Cancer Theraputicues*, *7*, 1789-1796.

Yuan, T., Nahar, P., Sharma, M., Liu, K., Slitt, A., Aisa, H. A. & Seeram, N. P. (2014). Indazole-type alkaloids from *Nigella sativa* seeds exhibit antihyperglycemic effects via AMPK activation *in vitro*. *Journal of Natural Products*, *77*, 2316–2320.

Yüncü, M., Şahin, M., Bayat, N. & İbrahim, S. (2013). Çörek otu yağının sıçan karaciğer gelişimine etkisi. *Gaziantep Medical Journal*, *19*, 180-184. [Effect of black seed oil on rat liver development.]

BIOGRAPHICAL SKETCH

Heba Ibrahim Mohamed

Affiliation: Ain Shams University

Research and Professional Experience: plant physiology, plant biochemistry, plant molecular biology, biotic and abiotic stress

Honors: Dr. Heba has long experience in research and obtained certificate of recognition in honor of achievement in international publication that supports Ain Shams University World Ranking 25/6/2012. Dr. Heba has obtained country encouragement prize in Cairo 2018.

Publications from the Last 3 Years:

[1] Mohamed, H. I., Latif, H. H. & Hanafy, R. S. (2016). Influence of nitric oxide Application on Some Biochemical Aspects, Endogenous Hormones, minerals and phenolic compounds of *Vicia faba* plant grown under arsenic stress. *Gesunde Pflanzen.*, *68*, 99-107. Impact factor 0.38.

[2] Helmi, A. & Mohamed, H. I. (2016). Biochemical and ulturasturctural changes of some tomato cultivars to infestation with *Aphis gossypii* Glover (*Hemiptera: Aphididae*) at Qalyubiya, Egypt. *Gesunde Pflanzen.*, *68*, 41–50. Impact factor 0.38.

[3] Latif, H. H. & Mohamed, H. I. (2016). Exogenous applications of moringa leaf extract effect on retrotransposon, ultrastructural and biochemical contents of common bean plants under environmental stresses. *South African Journal of Botany.*, *106*, 221-231. Impact factor 1.4.

[4] Mohamed, H. I., Mohammed, A. H. M. A. & Mogazy A. M. (2016). Effect of plant defense elicitors on soybean (*Glycine max* l.) growth, photosynthetic pigments, osmolyts and lipid components in response to cotton worm (*Spodoptera littoralis*) infestation. *Bangladesh Journal of Botany.*, *45*(3), 597-604. Impact factor 0.38.

[5] Mohamed, H. I., Elsherbiny, E. A. & Abdelhamid, M. T. (2016). Physiological and biochemical responses of *Vicia faba* plants to foliar application with zinc and iron. *Gesunde Pflanzen.*, *68*, 201-212. Impact factor 0.38.

[6] Akladious, S. A. & Mohamed, H. I. (2017). Physiological role of exogenous nitric oxide in improving performance, yield and some biochemical aspects of sunflower plant under zinc stress. *Acta Biologica Hungarica.*, *68*(1), 101–114. Impact factor 0.44.

[7] Mohamed, H. I. & Akladious, S. A. (2017). Changes in antioxidants potential, secondary metabolites and plant hormones induced by different fungicides treatment in cotton plants. *Pesticide Physiology and Biochemistry.*, *142*, 117-122. Impact factor 3.4.

[8] Aly, A. A., Mansour, M. T. M. & Mohamed, H. I. (2017). Association of increase in some biochemical components with flax resistance to powdery mildew. *Gesunde Pflanzen.*, *69* (1), 47-52. Impact factor 0.64.

[9] Mohamed, H. I. & Latif, H. H. (2017). Improving the tolerance of soybean genotypes to water stress by foliar application with methyl jasmonate. *Physiology and Molecular Biology of Plants.*, *23*(3), 545-556. Impact factor 1.15.

[10] Ashry, N. A., Ghonaim, M. M., Mohamed, H. I. & Mogazy, A. M. (2018). Physiological and molecular genetic studies on two elicitors for improving the tolerance of six Egyptian soybean cultivars to cotton leaf worm. *Plant Physiology and Biochemistry.*, *130*, 224-234. Impact factor 3.4.

[11] Mohamed, H. I., Akladious, S. A. & Ashry, N. A. (2018). Evaluation of six soybean genotypes using retro elements and some physiological parameters under water stress. *Gesunde Pflanzen.*, *70*, 205-215. Impact factor 0.79.

[12] Akladious, S. A. & Mohamed, H. I. (2018). Ameliorative effects of calcium nitrate and humic acid on the growth, yield component and biochemical attribute of pepper (*Capsicum annuum*) plants grown under salt stress. *Scientia Horticulturae.*, *236*, 244–250. Impact factor 1.96.

[13] Mohamed, H. I., Akladious, S. A. & El-Beltagi, H. S. (2018). Mitigation the harmful effect of salt stress on physiological, biochemical and anatomical traits by foliar spray with trehalose on wheat cultivars. *Fresenius Environmental Bulletin.*, *27*(10), 7054-7065. Impact factor 0.69.

[14] El-Beltagi, H. S., Mohamed, H. I., Safwat, G., Megahed, B. M. H. & Gamal, M. (2018). Evaluation of some chemical constituents, antioxidant, antibacterial and anticancer activities of *Beta vulgaris* L. root. *Fresenius Environmental Bulletin.*, *27*(9), 6369-6378. Impact factor 0.69.

[15] Mohamed, H. I., El-Beltagi, H. S., Aly, A. A. & Latif, H. H. (2018). The role of systemic and non systemic fungicides on the

physiological and biochemical parameters in *Gossypium hirsutum* plant, implications for defense responses. *Fresenius Environmental Bulletin.*, *27* (12), 8585-8593. Impact factor 0.69.

[16] El-Beltagi, H. S., Mohamed, H. I., Abdelazeem, A. S., Youssef, R. & Safwat, G. (2019). GC-MS analysis, antioxidant, antimicrobial and anticancer activities of extracts from *Ficus sycomorus* fruits and leaves. *Notulae Botanicae Horti Agrobotanici Cluj-Napoca.*, *47*(2), 493-505. Impact factor 0.65.

[17] El-Beltagi, H. S., Mohamed, H. I., Elmelegy, A. A., Eldesoky, S. E. & Safwat, G. (2019). Phytochemical screening, antimicrobial, antioxidant, anticancer activities and nutritional values of cactus (*Opuntia Ficus Indicia*) pulp and peel. *Fresenius Environmental Bulletin*, *28*(2A), 1534-1551. Impact factor 0.69.

[18] El-Beltagi, H. S., Mohamed, H. I., Safwat, G., Gamal, M. & Megahed, B. M. H. (2019). Chemical composition and biological activity of *Physalis peruviana* L. *Gesunde Pflanzen.*, *71*, 113–122. Impact factor 0.79.

[19] Mohamed, H. I., Ashry, N. A. & Ghonaim, M. M. (2019). Physiological analysis for heat shock induced biochemical (responsive) compounds and molecular characterizations of ESTs expressed for heat tolerance in some Egyptian maize hybrids. *Gesunde Pflanzen.*, *71*, 213-222. Impact factor 0.79.

[20] Manal, M. Hamed., Mohammed, A. Abd El-Mobdy., Marian, T. Kamel., Mohamed, H. I. & Abdelrahman, E. Bayoumi. (2019). Phytochemical and biological activities of two asteraceae plants *Senecio vulgaris* and *Pluchea dioscoridis* L. *PharmacologyOnline.*, *2*, 101-121.

[21] Adel, A Rezk., Jameel, M. Al-Khayri., Abdulaziz, M. Al-Bahrany., Hossam, S El-Beltagi. & Mohamed, H. I. (2019). X-ray irradiation changes germination and biochemical analysis of two genotypes of okra (*Hibiscus esculentus* L.). *Journal of Radiation Research and Applied Sciences*, *12*(1), 393–402. Impact factor 2.96.

[22] Mahmoud, T. M. Mansour., Aly, A. Aly., Marian, M. Habeb. & Heba I. Mohamed. (2020). *Control of cotton seedling damping-off by treating seed with inorganic salts*. Under publication.

[23] Mohamed, H. I., Youssef, S. A., Sedky, M. S. & Rafaat, H. (2020). *Use Polymerase Chain Reaction technique to identify Erwinia carotovora the causal agent of soft rot disease in potato*. Under publication.

[24] Marwa, M. Ghonaim., Ahmed, A. A. Omran. & Heba, I. Mohamed. *Evaluation of wheat salt stress tolerance using physiological parameters and retrotransposon-based markers*. Under publication.

In: *Nigella sativa*
Editor: Sanjin Berghuis

ISBN: 978-1-53617-538-7

Chapter 3

THERAPEUTIC POTENTIALS OF *NIGELLA SATIVA* MAJOR COMPONENT: THYMOQUINONE

Sunita Singh*[*]*, PhD
Department of Chemistry, Navyug Kanya Mahavidyalaya,
University of Lucknow, Lucknow, India

ABSTRACT

Recently the World Health Organization has urged and encouraged countries of the developing world to upgrade their primary healthcare programs with the incorporation of traditional medicinal plant therapy. One of the most studied medicinal plants is *Nigella sativa*. *N. sativa* from the family *Ranunculaceae* is an annual flowering plant also called black cumin, black seed, or Habbatul Barakah. It is also described as the "miracle herb" of the century. *N. sativa* is native to south and southwest Asia wherein the plant is cultivated and grows. It is also widely cultivated in Mediterranean countries, middle Europe and western Asia. It has amazing curative and therapeutic features that make them one of the most popular, safe, non-detrimental, and cytoprotective medicinal plant that

[*] Corresponding Author's E-mail :oceans.singh@gmail.com.

can be used for prevention and treatment of many complicated diseases. *N. sativa* seeds and oil extract have been known to be positively implicated in allergy, diabetes, cardiovascular problems like hypertension, hyperlipidemia, gastro- intestinal problems, inflammatory and oxidative damage processes. Moreover, its antihistaminic, antioxidative and immunomodulator properties cover a wide range of degenerative health problems apart from few recent reports of its anti-viral properties. *N. sativa* seeds contain diverse but well-characterized chemical components; which include both fixed and essential (volatile) oil, proteins and amino acids, carbohydrates, alkaloids, organic acids, saponins, crude fibers, vitamins, and minerals. Thymoquinone (2-isopropyl-5-methylbenzo-1, 4-quinone) (TQ), the most abundant constituent of the volatile oil also present in the fixed oil, is the biologically active component of *N.sativa* seeds. TQ as a bioactive component is found in many medicinal plants. Apart from *Ranunculaceae* family, the presence of this compound has been confirmed in several genera of the *Lamiaceae* family such as *Agastache*, *Coridothymus*, *Thymus*. It has also been found in genus *Teraclinis* and in the form of glycoside in Juniperus of *Cupressaceae* family. *N.sativa* seeds contain more than 100 chemical compounds, a number of them are yet to be characterized.

However, much of the biological activity of the *N. sativa* including antihypertensive nephroprotection, antipyretic, antimicrobial, and antineoplastic has been attributed to presence of TQ in it. The major pharmacological activities demonstrated by TQ are anticonvulsant, antimicrobial, anticancer, anti-histaminic, antidiabetic, anti-inflammatory, and antioxidant. More recently, a great deal of attention has been given to this dietary phytochemical with an increasing interest to to assess its health benefits through pre-clinical and clinical researches. Considering its wide range of medical applications TQ can be a potent candidate to be used as a natural drug. Considering the extraordinary medicinal attributes of TQ, this chapter accounts for therapeutic potentials of TQ.

Keywords: anticonvulsant, antimicrobial, anticancer, antihistaminic, antidiabetic, anti-inflammatory, and anti-oxidant

ABBREVIATIONS

TQ	Thymoquinone
SOD	superoxide Dismutase
SC-CO_2	supercritical CO_2

CAT	catalase
CTX	cyclophosphamide
GSH	reduced Glutathione
GSSG	oxidized glutathione
PPAR	Peroxisome proliferator-activated receptors
COX	Cycloxygenase
PG	Prostaglandin
NADPH	Nicotinamide adenine dinucleotide phosphate-oxidase
DPPH	1,1- diphenyl-2-picrylhydrazyl
MDA	malondialdehyde
NO	nitric oxide
PARP	Poly (ADP-ribose) polymerase
cdk	cyclin-dependent kinase
HO-1	Heme oxygenase-1
NF-kb	nuclear factor kappa B
TNF	Tumor necrosis factor
GTP	guanosine triphosphate
LPS	Lipopolysaccharides
IL-6	Interleukin 6
LDLC	low density lipoprotein cholesterol
LDLR	low density lipoprotein receptor
MICs	minimum inhibitory concentrations
DAPI	4,6-diamidino-2- phenylindole
GT	glutathione transferase
DHTQ	Dihydrothymoquinone
PTZ	pentylene tetrazole
TLR4	toll-like receptor 4
TBARS	Thiobarbituric acid reactive substances NPSH, non-protein sulfhydryl
NOX-4	renal oxidase. MDR. Multi-drug resistant
MCF-7	Michigan cancer foundation7
MiaPaCa-2	human pancreatic cancer cell line
HL-60	human leukemia cell line
GABA	Gamma amino butyric acid

Aβ	Beta amyloid peptides
Akt/PI3K	phosphoinositide-3-kinase–protein kinase B
PTEN	Phosphatase and Tensin Homolog
CHEK1	checkpoint kinase 1
DOX	doxorubicin
ROS	Reactive oxygen species
GSSG	Glutathione disulfide
Nrf	Nuclear repiratory factor
STZ	streptozotocin
cGMP	cyclic guanosine monophosphate
CCl-12	chemokine (C- C motif) ligand 12
Cxcl-10	chemokine (C-X-C motif) ligand 10
MCP	murine monocyte chemoattractant protein
GCSF	granulocyte stimulating factor
ALT	alanine aminotranaminase
LDH	lactate dehydrogenase
UHRF1	ubiquitin like containing PHD and RING finger domains

1. Introduction

With the great strides recently made in the area of optimum nutrition, nowadays there is a resurgence of interest in the use of plants as a source of food and medicine [1]. Recently, the usage of phytomedicine has been amplified dramatically for numerous ailments because of not only their easy accessibility and low cost but also the belief that natural remedies have fewer harmful effects when compared to synthetic medicines [2]. The development of new products from natural sources is also encouraged because it is estimated that, of the 300,000 herbal species that exist globally, only 15% have been explored for their pharmacological potential [3]. Among several medicinal plants, *Nigella sativa* L. (*Ranunculaceae*) has been considered one of the most treasured nutrient-rich herbs in history around the world and numerous scientific studies are in progress to validate the traditionally claimed uses of small seed of this species [4]. *N.*

sativa seeds, which are the rich source of the active ingredients of plant, have long been used in the Middle and Far East as a traditional medicine for a wide range of pathological conditions [5]. It is used in ethnomedicine to treat ailments and symptoms including, asthma, bronchitis, inflammation, eczema, fever, influenza, hypertension, cough, headache, dizziness, diabetes, kidney and liver dysfunctions, nervous disorders, rheumatism, cancer and related inflammatory diseases, gastrointestinal problems, and overall for general well-being [6]. *N. sativa* seeds contain diverse but well-characterized chemical components; they include both fixed and essential (volatile) oil, proteins and amino acids, carbohydrates, alkaloids, organic acids, saponins, crude fibers, vitamins, and minerals [7].

Thymoquinone (TQ) is the most abundant bioactive component of volatile oil of *N. sativa*. It was reported that the biological activities of *N. sativa* seeds are mainly ascribed to its essential oil constituent that is TQ (30–48%) and was first extracted by El–Dakhakhny [8]. *N. sativa* seed oil is cataloged in the list of United States Food and Drug Administration as "Generally Recognized as Safe." The major pharmacological activities exerted by TQ included anticonvulsant, anti-microbial, anti-cancer, anti-histaminic, antidiabetic, anti-inflammatory, and antioxidant [9]. Apart from *Rannunculaceae* family, the presence of TQ in several other genera of the *Lamiaceae* family including as *Agastache*, *Coridothymus*, *Monarda*, *Mosla*, *Origanum*, *Satureja*, and *Thymus* were also reported [10-13]. It has also been found in genus *Tetraclinis*, and in the form of glycoside in the genera *Cupressus* and *Juniperus* of the *Cupressaceae* family.

1.1. Phytochemistry of *N. Sativa*

Several bioactive compounds from the seed of *N. sativa* have been reported in the literature; among those the most important bioactive ones are thymoquinones. Other main phytochemicals reported from different varieties of *N. sativa* include sterols and saponins, phenolic compounds, alkaloids, novel lipid constituents and fatty acids, and volatile oils of varying composition [14]. The essential oil composition (0.4-0.45%)

reported in various studies represented about forty different compounds, amongst the abundantly constituents identified are *trans*-anethole, *p*-cymene, limonene, carvone, α-thujene, TQ, thymohydroquinone (THQ), dithymoquinone, carvacrol, and β- pinene with various concentration [15]. The quantity of most important bioactive constituent, TQ, present in the volatile oil isolated by different extraction methods from the seeds of *N. sativa* varied over a wide range: using SC-CO_2 (1.06, 4.07mg/g) [16] and by Soxhlet extraction (2940.43mg/kg) [17] and (8.8mg/g) oil [18].

The seed oil fatty acid composition (32-40%) has been reported by various authors to contain mainly, linoleic, linolenic, oleic, palmitoleic, palmitic acids together with arachidonic, eicosadienoic, stearic, and myristic acid [19]. A new dienoate and two known monoesters along with novel lipids have been isolated from the unsaponified extract of the seed, namely methylnonadeca-15,17-dienoate, pentyl hexadec-12-enoate, and pentyl pentadec-11-enoate [20]. Phytosterols are important part of human diet and are gaining greater interest due to their nutraceutical and medicinal benefits in lowering low density lipoprotein and total cholesterol level [21]. Phytosterols are also important as characteristic compounds for assessing the quality of vegetable oils and food labeling. The total sterols content of black cumin seed oil as estimated by different researchers was found to be between 18 and 42% of the unsaponified matter. The major sterols identified were β-sitosterol, campesterol, stigmasterol, and 5-avenasterol [22]. Tocopherols exhibited attractive scavenging potentials of free radicals which are believed to terminate lipids peroxidation [23]. The total tocopherol contents of black seed oil reported in varied quantities from diverse sources ranged from 9.15 to 27.92mg/100 g. Among the foremost tocopherols recognized in black cumin seeds, α- and γ-tocopherol and β-tocotrienol are well recognized [19]. Moreover, alkaloids of diverse types have been isolated from the seeds of black cumin, which include novel Dolabellane-type diterpene alkaloids: nigellamines A1, A2, B1, and B2 and nigellamines A3, A4, A5, and C [24] possessing lipid metabolizing property, and imidazole class of alkaloids: nigellidine, nigellicine [25], and nigellidine-4-O-sulfite [26].

1.2. Pharmacokinetics and Physicochemical Properties of TQ

Different dose regime of TQ which were administered either by intraperitoneal or intravenous or intragastric routes were studied for their efficacy in diseases models. The experimental studies in animals reveal the usual tested therapeutic doses of TQ 5mg/kg (intravenous) and 20 mg/kg per oral [27]. The pharmacokinetic parameters were calculated using high performance chromatography (HPLC) in the collected blood samples from vole rabbits. The mobile phase contains methanol and potassium dihydrogen phosphate buffer and carried at the rate of 0.9 ml/min. The compound, TQ was detected at the wavelength of 254 nm. The estimated clearance (CL) following IV administration was found to be 7.19 ± 0.83 ml/kg/min, and volume of distribution at steady state (Vss) was 700.90 ± 55.01 ml/kg. While, subsequent oral dosing, the CL/F and Vss/F-value were found to be 12.30 ± 0.30 ml/min/kg and 5,109.46 ± 196.08 ml/kg, respectively. These parameters remained connected through elimination half-life (t1/2) of 63.43 ± 10.69 and 274.61 ± 8.48 min with intravenous and oral administration. The proposed absorption $t_{1/2}$ was about 217 min. The compartmental analysis revealed $t_{1/2a}$ of *8.9 min and $t_{1/2b}$ of *86.6 min. The projected complete bioavailability of TQ stayed *58%through a lag time of *23min [27]. The protein binding of TQ was found 99% with a composite rapid abolition and moderately gentler absorption following oral administration. TQ was highly sensitive to light, even after a short period of exposure. A short period of light expo-sure led to severe degradation, independent of the solution pH and solvent type. In addition, it was unstable in aqueous solutions, particularly at an alkaline pH. TQ stability decreased with rising pH; at alkaline pH, it suffered the highest degradation rate with the minimal degradation being at acidic pH [28].

The solubility of TQ molecule is a challenge due to its hydrophobicity which direcly affects its bioavailability and cause limitations in drug formulation. In addition, its solubility varies with time and ranges from 549 to 669 g/ml in aqueous solutions at 24 h, increasing at 72 h to 665–740g/ml [28]. Therefore, recently attempts have been made to synthesize novel analogs of TQ with enhanced bioavailability and activity. Numerous studies have been involved to develop the analogs of TQ with remarkable efficacy for different types of cancers. Recently, the novel analogs of TQ were analyzed with various monoterpenes, sesquiterepene, and cytotoxic terpenes for better anticancer activity. These analogs were tested against various cancer cell lines to demonstrate cell specific activity. The resulting data proposed that some of these analogs were more potent than TQ, the parent drug at the same time being less cytotoxic to the non-malignant cells.

The lack of bioavailability and pharmacokinetic parameters, and formulation problems delayed the usage of TQ in clinical phase. Hence, more investigation is needed for a better understanding of TQ pharmacological properties meant for future clinical develop-ment. Taken together, the data demonstrate that TQ has enormous potential for its clinical usage and pharmaceutical development. So, many new strategies of targeted drug delivery and newer techniques were developed including nanotechnology.

2. Methods

Databases such as PubMed, Science Direct, Scopus, and Google Scholar were searched for the keywords like of *N. sativa*, TQ, therapeutic potentials, pharmacological properties and different disorders between the years 1979 and 2017 to prepare this chapter. For validating the plant's scientific name, Plantlist.org and examine.com were used.

3. Therapeutic Potentials of TQ

Therapeutic potentials of TQ (Figure 1.) were discussed in this section which is definitely due to the presence of pharmacological properties of TQ. *N. sativa* has been broadly studied in the last few decades and studies have reported that it possesses a number of medicinal properties and pharmacological actions. Different pharmacological properties were discussed in this section.

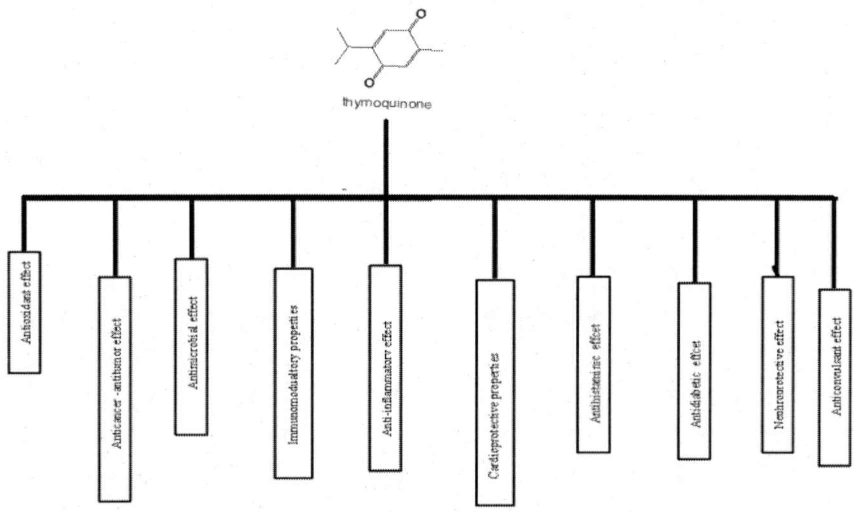

Figure 1. Structure of Thymoquinone and its different therapeutic properties

3.1. Antioxidant Activity

Aerobic metabolism as well as UV radiation and environmental pollution causes the persistent generation of superoxide anion radical ($O^{2-\bullet}$), hydroxyl radical (•OH) and peroxyl radical (ROO•). The sustained elevated levels of reactive oxygen species (ROS) cause oxidation of biomolecules including lipid membrane, protein and nucleic acids. TQ possesses free radical scavenging activities against free radicals which initiate oxidative stress in various *in vitro* and *in vivo* animal models. Due to potent antioxidant and free radical scavenging action, TQ has been shown to normalize the adverse effect of various environmental toxins or xenobiotics causing oxidative damage and organ dysfunctions and leads pathogenesis of various diseases [29]. TQ also impede the generation of oxidative stress by countering xanthine/xanthine oxidase system [30]. Figure. 2 shows the molecular targets of TQ during different oxidative processes [30]. SOD shows a vital role in altering superoxide anions into hydrogen peroxide and oxygen; hence it forms the first line anti-oxidant defense and known as an important detoxification enzyme. TQ has been shown to enhance the activity of superoxide dismutase (SOD). Glycation of SOD upon *in vitro* incubation of the SOD with glucose or methylglyoxal results in decreased activity of SOD that is an additional mechanism of TQ in modulating SOD activity and subsequent retention of the anti-oxidant activity [32]. TQ also showed the scavenging of the carbon-centered radicals and hydroxyl radicals as assessed in the *in vitro* assays; 1,1-diphenyl-2 picrylhydrazyl (DPPH) and iron-catalyzed injury of deoxyribose. Further, it also curtailed the development of oxidative stress by non-enzymatic interface involving the glutathione (GSH) cycle to produce glutathionylated-dihydrothymoquinone [33]. Both the reduced metabolites of TQ exhibited strong free radical hunting action against 2,2-azinobis (3- ethylbenzothiazoline-6-sulfonic acid) and DPPH assay by giving hydrogen to free radicals. Since lipid peroxidation products form adduct through proteins or DNA, the inhibitory effect of TQ on lipid peroxidation in part accounted for its beneficial effects. Altogether, the studies demonstrate that TQ attenuate cellular oxidative stress by inducing GSH under different experimental pathologic conditions [34]. Convincing data demonstrate that the uncontrolled generation of free radicals target the lipid membranes and induce a state of anti-oxidant defense that leads the initiation of lipid peroxidation. The occurrence of lipid peroxidation is an important pathogenic event in many diseases and can be measured by the biomarkers of lipid peroxidation *in vitro* and *in vivo* models of human diseases. Until now, convincing number of experimental studies have recommended that TQ provide protection against many chemicals and drugs such as ifosfamide, mercuric chloride, cyclophosphamide, isoproterenol, cisplatin, and doxorubicin by augmenting anti-oxidant antioxidant defense and attenuating inflammation and apoptosis.

Pretreatment of Wistar rats with TQ and 1,2-dimethylhydrazine (DMH) for 10 weeks prevented the depletion of antioxidant enzymes catalase, glutathione peroxidase, and superoxide dismutase in red blood cells and maintained a similar value as the control group. At the same time, it prevented erythrocyte damage in DMH-induced colon post initiation carcinogenesis in rats [35].

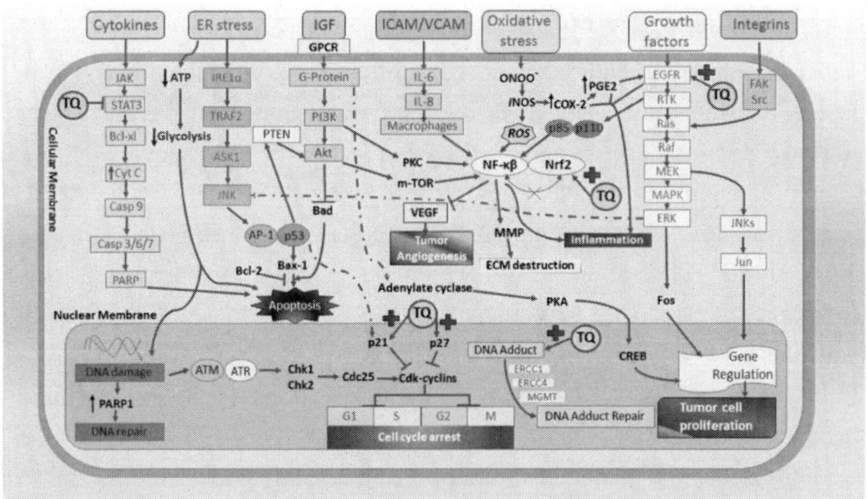

Figure 2. Different molecular targets of TQ during oxidative processes.

TQ and *N. sativa* oil possess cytoprotective effects against the anti-cancer drugs cyclophosphamide (CTX) via maintenance of hemoglobin and blood sugar levels, and the activities of liver enzymes, bilirubin, urea, creatinine, lipids (triglyceride, cholesterol and low density lipoprotein (LDL)-cholesterol) and lipid peroxidation in the liver. The cytoprotective effects of *N. sativa* oil and TQ were associated with induction of antioxidant mechanisms [36]. Neuron-protective effects have also been studied in cultured hippocampal and cortical neurons treated with amyloid-β peptide (Aβ1-42) and TQ simultaneously for 72 h. TQ efficiently attenuated Aβ1-42-induced neurotoxicity by improving cell viability. It has also been shown to inhibit mitochondrial membrane potential depolarization and the generation of reactive oxygen species caused by Aβ1-42, and to restore synaptic vesicle recycling inhibition and to partially reverse the loss of spontaneous firing activity, and Aβ1-42 aggregation *in vitro* [37].

3.2. Anti-Inflammatory Properties

There are many reports on the anti-inflammatory activity of TQ [38]. Kundu and coworkers [39], stated that the anti-inflammatory effect of TQ is caused by the upregulated expression of heme-oxygenase 1 (HO-1) in human keratinocytes (HaCaT) by activating nuclear factor (NF)-erythroid2-(E2)-related factor-2 (Nrf2) via reactive oxygen species (ROS)-mediated phosphorylation of protein kinase B (PKB/Akt) and cyclic AMP-activated protein kinase-alpha (AMPK alpha). According to Bai and coworkers [40], TQ attenuated thioacetamide (TAA)-induced liver fibrosis accompanied by reduced protein and mRNA expression of of α-smooth muscle actin (α-SMA), collagen-I and tissue inhibitor of toll-like receptor 4(TLR4) and decreased pro-inflammatory cytokine levels. It also inhibited phosphatidylinositol 3-kinase phosphorylation and enhanced the phosphorylation of adenosine monophosphate-activated protein kinase (AMPK) and liver kinase B (LKB). TQ has also been reported to inhibit the effects of 12-O-tetradecanoylphorbol-13-acetate (TPA)-induced expression of cyclooxygenase-2 (COX-2) and nuclear factor kappa-light-chain-enhancer of activated B cells (NF-κB) (41). *N. sativa* and TQ treatment also suppressed the expression of the COX-2 enzyme in the pancreatic tissue of streptozotocin (STZ)-induced diabetic rats [42]. The antiulcerative effect of *N. sativa* and TQ was demonstrated by Kanter and coworkers [43] by investigating ethanol induced mucosal ulceration in rats, which was inhibited by pretreatment with TQ and *N. sativa*. Furthermore, oral administration of TQ in Wistar rats at 5mg/kg body weight for 21 days led to a significant reduction of the levels of different pro-inflammatory mediators (IL-1β, IL-6, TNFα, IFNγ and PGE(2)) [44]. Intraperitoneal treatment of mice with TQ (6 mg/kg; IP), 24 and 1 hr before intratracheal treatment with Diesel exhaust particles (DEP) (30 μg/mouse), prevented pulmonary inflammation and the increase of airway resistance caused by DEP, and inhibited the increase of blood leukocyte numbers and plasma IL-6 concentrations [45]. The effects of TQ on airway inflammation in a mouse model of allergic asthma were investigated by intraperitoneal injection of TQ before airway challenge of ovalbumin (OVA)-sensitized mice, and caused a marked decrease in lung eosinophilia and elevated Th2 cytokines - both *in vivo* and *in vitro* - following stimulation of lung cells with OVA. TQ also decreased the elevated serum levels of OVA-specific IgE and IgG1. Histological examination of lung tissue demonstrated that the compound significantly inhibited allergen-induced lung eosinophilic inflammation and mucus-producing goblet cells [46]. Using an asthmatic murine model, TQ has also been demonstrated to have a high potential in inhibiting the inflammatory changes associated with asthma, especially the aggregation of inflammatory cells in bronchoalveolar lavage (BAL) fluid and in lung tissues. In addition it inhibited mRNA expression of inducible nitric oxide synthase (iNOS) and transforming growth factor-β1 (TGF-β1) [47]. In experiments on ovalbumin-sensitized guinea pigs and sulfur mustard exposed guinea pigs, an outstanding evidence of the preventive anti-inflammatory effects of TQ and *N. sativa* has been reported [48]. Different extracts, mainly aqueous extracts, from *N. sativa* seeds proved to possess relaxant (bronchodilatory) effects on tracheal chains of guinea pigs [49]. Taken together, these studies have collectively supported the notion that TQ could be useful in intervening the inflammatory cascade, which may lead to the inhibition of cancer progression and thus improve patients' morbidity and mortality

3.3. Immunomodulatory Properties

Experimental studies have reported that TQ reduced TNF-α, interleuken-6 and IL-1β in blood and tissues, and protected tissues by reducing inflammation [50]. TQ inhibited IL-6 signaling [51]; and it is possible that the role of TQ in EAE [52], rheumatoid arthritis [53], allergic asthma [46], allergic lung inflammation [38], experimental colitis [44], and immunomodulatory effects [5] is due to suppression of IL-6 signaling. Furthermore, TQ suppressed NO overproduction and iNOS expression induced by different agents in various animal tissues [54]. It diminished iNOS levels in LPS-activated rat peritoneal macrophages and thus suppressed the production of NO by macrophages [55], an effect that may be useful in ameliorating the inflammatory and autoimmune conditions. TQ also acted *in vitro* as inhibitor of COX-1and -2 [56], also inhibiting p38 in murine macrophages [57]. The immunomodulatory effect of TQ was observed also in mixed lymphocyte cultures, where it caused decreased secretion in the levels of the cytokines IL-1β and -8 [58]. It enhanced the survival and activity of antigen-specific CD8T cells thus protecting against tumor growth [59]. TQ might regulate immunity also by influencing dendritic cell functions such as maturation, cell pH (by affecting Na+/H+ activity), oxidative burst, migration and cytokine release and survival. Dendritic cell volume may also be affected by TQ [60]. TQ may enhance the activation of Nrf2, thereby raising the expression of heme oxygenase-1 (HO-1). TQ induced HO-1 expression in HaCaT cells by activating Nrf2 through ROS mediated phosphorylation of Akt and adenosine monophosphate-activated protein kinase (AMPK) as upstream targets [39].

3.4. Anticancer and Anti-Tumor Activity

The research area of anti-neoplastic drugs is currently under investigation, products of natural origin have gained considerable attention. Natural compounds and phytochemicals may be exceptional resource for anti-cancer agents, and may be valuable alternatives to synthetic drugs in cancer therapy, or used to enhance their effect and reduce their dosing and limiting their toxicity [61]. An ideal cancer therapeutic agent is one that exerts its anti-cancer effect with partial cytotoxicity over normal tissues. One of the advantages of TQ is that its anti-cancer effects has been shown to be activated more specifically against cancer cells than normal cells [62]. As detailed below, among a wide spectrum of phytochemicals TQ is a compound exhibiting promising anti-carcinogenic, anti-neoplastic, anti-proliferative and anti-mutagenic activities against miscellaneous tumor models. It also is chemopreventive, and reduces the toxic effects of standard anti-neoplastic agents, as well was a chemosensitizer when used in combination with chemotherapeutic agents, while it was minimally toxic to normal cells. TQ was shown to have a dual role, and depending on the cellular microenvironment, it may act as an anti-oxidant or as a pro-oxidant. It exerted its biological functions by ROS generation in tumor cells where it acted as pro-oxidant. TQ reacted with amino or thiol groups of amino acids, undergone a series of oxido-reduction reactions, and then metabolized to semiquinone or thymohydroquinone, ultimately leading to the production of ROS [63].

There is evidence that TQ induces p53-independent apoptosis via the activation of caspase-8 and caspases 9 and 3 in the caspase cascade. Activation of caspase-8 promotes release of cytochrome *c* from mitochondria into the cytoplasm. It also modulates the Bax/Bcl2 ratio by up regulation of proapoptotic Bax and down-regulation of antiapoptotic Bcl2 proteins in p53-null HL-60 cells during apoptosis [64]. Investigating the anti-cancer effects of TQ on A549 non-small cell lung cancer cells exposed to benzo(a)pyrene, Ulasli and coworkers [65] found that TQ treatment up-regulated Bax and down regulated Bcl2 proteins, and increased the Bax/Bcl2 ratio. It also decreased the expression of cyclin D and increased the expression of p21, and it up-regulated TRAIL receptor 1 and 2 expression. These molecular events lead to regulatory p53 levels affecting the induction of G2/M cell cycle arrest and apoptosis.

In breast cancer cells TQ was able to increase peroxisome proliferator-activated receptor gamma (PPAR-γ) activity and to down-regulate the expression of the genes for Bcl-2, Bcl-xL and survivin. More importantly, the increase in PPAR-γ activity was prevented in the presence of PPAR-γ specific inhibitors and PPAR-γ dominant negative plasmids, suggesting that TQ may act as a ligand of PPAR-γ [66]. Treatment of human breast carcinoma in both *in vitro* and *in vivo* models demonstrated antiproliferative and proapoptotic effects of TQ, which are mediated by its inductive effect on p38 and ROS signaling. TQ possesses anti-tumor effects in breast tumor xenograft mice and it potentiates the antitumor effect of doxorubicin [67]. TQ has also been shown to inhibit the growth of the human cholangiocarcinoma (CCA) cell lines TFK-1 and HuCCT1 in a dose- and time-dependent manner. Themechanism of CCA cell line growth inhibition is exerted by down-regulation of PI3K/Akt and NF-κB, and regulated gene products, including X-linked inhibitor of apoptosis protein

(XIAP), vascular endothelial growth factor (VEGF), p-AKT, p65, Bcl-2 and COX-2 [69]. TQ also exerts an inhibitory effect on migration of metastatic human (A375) and mouse (B16F10) melanoma cells by inhibition of NLRP3 inflammasome resulting in a decreased proteolytic cleavage of caspase-1. Thus, it can be a potential immunotherapeutic agent not only in adjuvant therapy for melanoma, but also in the control and prevention of metastatic melanoma [70]. TQ is also a microtubule-targeting agent (MTA), and binds to the tubulin-microtubule network, thus preventing microtubule polymerization and causing mitotic arrest and apoptosis of A549 cells but not of normal HUVEC cells [71]. Investigating the putative anti-cancer activities of TQ on α/β tubulin expression in human astrocytoma cells (cell line U87, solid tumor model) and in Jurkat cells (T lymphoblastic leukaemia cells) evidence was provided for TQ to target the level of α/β tubulin proteins in cancer cells. It induced α/β tubulin in both cancer cell types. The degradation found was associated with the upregulation of the tumor suppressor p73 with subsequent induction of apoptosis. No effect on α/β tubulin protein expression was found in normal human fibroblasts used as control cell model. These data indicate that TQ exerts a selective effect on α/β tubulin in cancer cells [72]. Furthermore, TQ effects on human topoisomerase IIα were investigated and demonstrated that it enhances enzyme-mediated DNA cleavage 5-fold, which is similar to the anti-cancer drug etoposide indicating that TQ can be considered as human type II topoisomerase poison [73]. The majority of patients with glioblastoma, the most aggressive malignant astrocytic brain tumor in adults, experience a recurrence of the tumor because of these cells` resistance to apoptotic cell death following ionizing radiation and chemotherapy with temozolomide (TMZ), and an increased autophagy, TQ proved to induce caspase-dependent apoptosis and to inhibit autophagy of glioblastoma cells [74]. By studying the mechanisms of cytotoxicity on neuroblastoma (Neuro-2a) cells it was additionally found that TQ induces apoptosis by increasing the Bax/Bcl-2 ratio, which leads to the release of cytochrome c from mitochondria into the cytoplasm. TQ treatment also directs the activation of caspase-3 followed by the cleavage of poly (ADP-ribose) polymerase (PARP) and down regulates the caspase inhibitor XIAP [75].

Gali-Muhtasib and coworkers [76] evaluated the therapeutic potential of TQ in two different murine colon cancer models, viz. 1, 2-dimethyl hydrazine (DMH), and xenografts model. In the DMH model, TQ was injected intraperitoneally and the multiplicity, size, and distribution of aberrant crypt foci (ACF) and tumors were determined at Weeks 10, 20, and 30. TQ significantly reduced the numbers and sizes of ACF by 86%, and tumor multiplicity at Week 20 was reduced from 17.8 in the DMH group to 4.2 in mice injected with TQ. This effect persisted, and tumors did not regrow even when TQ injection was discontinued for 10 wk; and immunostaining for caspase 3 cleavage in remnant tumors confirmed increased apoptosis in response to TQ.

The growth inhibitory and antitumor effects of TQ were further studied by Badary and Gamal El Din [77] in fibrosarcoma induced by 20-methylcholanthrene (MC) in male Swiss albino mice. TQ was found effective not only in significantly inhibiting tumor incidence and tumor burden (34% compared to 100% in control tumor-bearing mice), but it also delayed the onset of MC-induced fibrosarcoma tumors—indicative of chemopreventive action against MC-induced fibrosarcomas.

Badary and coworkers [78] considered possible augmentation of the antitumor activity of cisplatin by TQ in Ehrlich ascites carcinoma (EAC)-bearing mice and authenticated that TQ (50 mg/l in drinking water), when given 5 days before and 5 days after single injection of cisplatin, abrogated cisplatin nephrotoxicity and potentiated the antitumor activity of cisplatin. Another study reported by Badary [79] in mice bearing EAC xenograft, documented that TQ (10 mg/kg/day) in drinking water significantly enhanced the antitumor effect of Ifosfamide. Mice treated with Ifosfamide in combination with TQ showed less body weight loss and mortality rate compared to Ifosfamide monotherapy. These observations demonstrate that TQ may improve the therapeutic efficacy of Ifosfamide and cisplatin and in addition, reverses Ifosfamide- and cisplatin-induced nephrotoxicity by preventing renal GST depletion and lipid peroxide generation and improving their antitumor efficacy.

Kaseb and coworkers [80] observed in a xenograft prostate tumor model that TQ inhibited growth of C4–2B derived tumors in nude mice. This was associated with a dramatic decrease in androgen receptor, transcription factor E2F-1, and cyclin A as determined by Western blot analysis. Their findings clearly suggest that TQ may prove to be an effective agent in treating hormone sensitive, as well as hormone refractory, prostate cancers with reasonable degree of selectivity. TQ was also shown in another study of human prostate cancer (PC3 cells) xenograft to inhibit the tumor growth and block angiogenesis with almost no toxic side effects [81].

Cytotoxicity of TQ was also tested in triple-negative breast cancer (TNBC) cells that lack functional tumor suppressor p53. TQ treated cells showed G1 phase cell cycle arrest and apoptosis characterized by the loss of mitochondrial membrane integrity as evidenced by release of cytochrome c and caspase 9 activation [82]. TQ treatment also inhibits the proliferation of multiple myeloma (MM) cells and potentiates the apoptotic effect of bortezomib in various MM cell lines via the activation of caspase-3, resulting in the cleavage of PARP. TQ treatment also inhibits chemotaxis and invasion induced by C-X-C motif chemokine 12 (CXCL12) in MM cells *in vitro* and a xenograft mouse model [82]. TQ treatment inhibits the expression of NF-κB and suppresses IL-8 and its receptors. It increases levels of ROS and mRNAs of the oxidative stress-related genes, NQO1 and HO-1. Pretreatment of HepG2 cells with N-acetylcysteine, a scavenger of ROS, prevented TQ-induced cell death. TQ treatment also stimulated mRNA expression of pro-apoptotic Bcl-xS and TRAIL death receptors, and inhibited expression of the anti-apoptotic gene Bcl-2. Conclusively, TQ enhanced TRAIL-induced death of HepG2 cells, in part by upregulating TRAIL death receptors, inhibiting NF-κB and IL-8 and stimulating apoptosis. These manifold molecular mechanisms of TQ-dependent suppression of HCC cell growth underscore the potential of this compound as anti-HCC drug [83].

3.5. Antimicrobial Effects

Kouidhi and coworkers [84] demonstrated the antimicrobial action of TQ, tetracycline and benzalkonium chloride using the efflux assay of 4,6-diamidino-2- phenylindole (DAPI) by determining DAPI cell accumulation. TQ provokes a 4-fold potentiation of the verified anti-biotics and antiseptics. TQ also reserved DAPI efflux action evidenced by increased accumulation in clinical isolates and decreased loss from bacteria that indicates the resistance-modifying action of TQ on oral bacteria. Additionally, the antimicrobial activity of *N. sativa* against various strains of *Staphylococcus mitis*, *S. mutans*, *S. constellatus*, and *G. haemolysins* were also demonstrated following the administration of essential oil containing 3.35µg of TQ. The minimum inhibitory concentration of TQ was found to be 19.25 ± 1.6 mg/ml. The study also revealed that TQ exerted potent inhibitory effect on Hep-2 cells as compared to the essential oil delineating its anti-cariogenic activity against bacterial infections [84-85]. The result of different studies [86] was carried out on 11 human pathogenic bacteria such as Gram positive cocci, *Staphylococcus aureus* ATCC25923 (22µg/ml) and *Staphylococcus epidermidis* CIP106510 (60µg/ml). In another study, retaining a sequence of trials, TQ was vigorous in contradiction of *Escherichia coli*, *Pseudomonas aeruginosa*, *Shigella flexneri*, *Salmonella typhimurium*, *Salmonella enteritidis*, and *S. aureus*. The concentration of TQ prerequisite to constrain and slay *S. aureus* was 400 and 800µg/ml. The studies suggest that use of TQ and dihydrothymoquinone with currently available modern antibiotics such as ampicillin, cephalexin, chloramphenicol, tetracycline, and gentamicin may provide synergistic activity in against *S. aureus* [87].

The antifungal activities of *N. sativa* oil, TQ, and griseofulvin were compared and it was found that the oil and its component, TQ showed greater minimum inhibitory concentration than griseofulvin against these fungi [88].

Table 1. Antibacterial effects of TQ

Treatment	**Method**	***Microorgansim***	**Results**
Aqueous extract (6.6 ml/kg once daily for 3 days) Aqueous extract showed inhibitory effect against candidiasis. Ether extract (2.5 – 40 mg/ml) TQ (0.062-.5 mg/ml)	Candidiasis in mice Agar diffusion method	*Candida albicans* Four species of *Trichophyton rubrum* and one each of *Trichophyton interdigitale*, *Trichophyton mentagrophytes*, *Epidermophyton floccosum* and *Microsporum canis*	Aqueous extract showed inhibitory effect against candidiasis. The MICs of the ether extract of *Nigella sativa* and TQ were between 10 and 40 and 0.125 and 0.25 mg/ml, respectively
Oil (1, 2 and 3 ml/ 100 ml)	Alfa Test-P affinity column	On the growth and aflatoxin B1 production by *Aspergillus parasiticus* (CBS 921.7) and *Aspergillus flavus* (SQU 21) strains	The inhibition of aflatoxin B1 production by *Aspergillus flavus* and *Aspergillus parasiticus* strains with different concentrations of *N. sativa* oil was in the range of 49.7- 58.3% and 32-48% respectively, but different concentrations of *N. sativa* oil displayed no significant effect on the growth of *Aspergillus* species
Oil (0.35% (v/w) yield)	Disc diffusion method and microdilution method	*Chrysosporium evolceanui Chrysosporium tropicum, Trichophyton simii, Trichophyton rubrum and Microsporum gypseum*	The maximum zone of inhibition was seen in *Micro. gypseum* with diameter of inhibition zone and activity index 38 mm (AI: 1.90), and the inhibition against *Tricho. rubrum*, *Tricho. simii* were (IZ: 20 mm, AI: 1.33 and IZ:35 mm, AI: 1.09), respectively. For *Chrysos. tropicum* was (IZ: 26 mm, AI: 0.86) and against *Chrysos. evolceanui* was (IZ: 25 mm, AI: 0.71)

The anti-rhinosinusitis potential of TQ was showed against rhinosinusitis induced by intranasal administration of platelet-activating factor TQ showed promising potential in the treatment of rhinosinusitis and found comparable to the standard anti-biotics. The results were further supported by the histopathological observations [89]. In additional to anti-bacterial and anti-fungal activities, TQ was also found effective against *Entamoeba histolytica* and *Giardia lamblia*, the common parasites causing amoebiasis and giardiasis [90].

Table 2. Antifungal effects of TQ

Treatment	**Method**	***Microorgansim***	**Results**
Diethyl ether extract (25-400 μ g/disc)	Filter paper discs impreg-nated	Gram-positive bacteria, Gram-negative bacteria and *Candida albicans*	Effective against Gram-positive (*Staphylococcus aureus*), Gram-negative bacteria (*Pseudomonas aeruginosa* and *Escherichia coli*), *C. albicans*, (not effective on *Salmonella typhimurium*), also effective against staphylococcal infection in mice
TQ and THQ	Disc diffusion	*Esch. coli, Pseudo. aeruginosa, Shigella flexneri, Sal. typhimurium, Salmonella enteritidis and Staph. aureus*	In the case of *Staph. aureus* the MIC and MBC for TQ were 3 and 6 μg/ml respectively, the MIC and MBC for THQ were 400 and 800 μg/ml respectively Gram-negative bacteria were less susceptible to both TQ and THQ and their MIC and MBC variety between 200 and 1600 μg/ml
TQ (0 to 512 μ g/ml)	Broth microdilution	Gram-negative bacilli: *Esch. coli* ATCC 35218, *Salmonella enteric* serovar *typhimurium* ATCC 14028, *Pseudo. aeruginosa* ATCC 27853, *Vibrio lginolyticus* ATCC 33787, *Vibrio paraheamolyticus* ATCC 17802; Gram-positive bacilli: *Bacillus cereus* ATCC 14579, *Listeria monocytogene* ATCC 19115 and Gram-positive cocci: *Enterococcus faecalis* ATCC 29212, *Micrococcus luteus* NCIMB 8166, *Staph. aureus* ATCC 25923, *Staphylococcus epidermidis* CIP 106510	TQ showed a significant bactericidal activity against the majority of the tested bacteria (MICs values ranged between 8 to 32 μg/ml), the best effect was seen especially in Gram-positive cocci (*Staph. aureus* ATCC 25923 and *Staph. epidermidis* CIP 106510)
Ethanolic extract (4 mg/disc)	Disc diffusion and agar dilution	Methicillin resistant *Staph. aureus* (MRSA)	All tested strains of MRSA were sensitive to extract and the extract had an MIC range of 0.2-0.5 mg/ml

TQ administered with curcumin shows anti-microbial action by elevating anti-body titer against H9N2 AIV in turkeys. The elevated cytokine gene countenance proposes anti-viral activity of TQ, curcumin and the combination synergistically repressed the pathogenesis of H9N2 viruses [91]. Selected studies showing the different extracts and microorganisms tested in experimental models *in vivo* and *in vitro* for antibacterial, antifungal also demonstrated in Tables 1 and 2 respectively [92].

3.6. Antihistaminic Effect

Histamine is released by basophils and mast cells, producing allergic reactions. Some of anti-inflammatory effects of TQ have been related to its role in inhibition of histamine production and/or release. Pretreatment with TQ inhibited the elicited increase in histamine release compared to the sensitized group. These results clearly indicated that TQ inhibited mast cell-mediated immediate type allergic reactions. It was reported that rat pretreated with *N. sativa* oil before induction of ulcer induced a significant decrease in gastric mucosal histamine content [93]. Pretreatment of the sensitized guinea pigs with TQ showed a significant decrease in the response of the tracheal spirals to histamine and acetylcholine compared to that produced with sensitized animals. The traditional use of *N. sativa* seeds and its active ingredients have a substantial impact on the inflammatory diseases mediated by histamine [94]. These results were supported by previous studies that investigated thymoquinones effect on the guinea pig isolated tracheal zig-zag preparation pre-contracted by carbachol. TQ caused a concentration-dependent decrease in the tension of the tracheal smooth muscle. TQ induced relaxation is probably mediated, at least in part, by inhibition of lipoxygenase (LO) products of arachidonic acid metabolism and possibly by non-selective blocking of the histamine and serotonin receptors. This relaxant effect of TQ, further supports the traditional use of black seeds to treat bronchial asthma [95].

3.7. Anticonvulsant and Neuroprotective Effect

Recent studies have been focused on the natural neuroprotective agents due to its low adverse effects with the increase in neurodegenerative diseases. Polyphenols have been considered as the main target for drug design due to the growing evidence that suggests that flavonoids possess beneficial effects on mental diseases. TQ is an important natural neuroprotective agent [96] that is widely seen in *N sativa* seeds. Many studies indicated the protective effects of TQ in the control of depression, epilepsy, Parkinson's disease, Alzheimer disease, ischemia, trauma brain injury, anxiety, encephalomyelitis, and brain cancer that have been found in many experiments and a few clinical trials. The results of different researches suggest an involvement of NO-cGMP and GABAergic pathways in the anxiolytic-like activity of TQ [97]. TQ also has potential to protect primary dopaminergic neurons against MPP (+) and rotenone relevant to parkinson's disease [98]. TQ pretreatment could attenuate seizure activity and lipid peroxidation, lower hippocampal neuronal loss, and mitigate astrogliosis in epilepsy model [99]. TQ may prevent neurotoxicity and $A\beta_{1-40}$-induced apoptosis [100]. The neuroprotective effects of TQ may be related to modulatory effects on inflammation, apoptosis, and oxidative stress [101]. The activation of the Nrf2/ARE signaling pathway by TQ resulted in the inhibition of NF-κB-mediated neuroinflammation. TQ exhibited anti-inflammatory effects by decreasing several cytokines, including TNF-α, NF-κB, IL-6, IL-1β, IL-12p40/70, (CCL12)/MCP-5, (CCL2)/MCP-1, GCSF, and Cxcl 10/IP-10 of, NO, PGE2, and iNOS [102]. TQ modulates oxidant–antioxidant system by increasing antioxidant content, including GSH, CAT, glutathione *S*-transferase, and SOD, and decreasing lipid peroxidation in brain tissue [103]. The antianxiety effects of TQ may be related to the modulating effects on NO-cGMP and GABA-ergic pathways [97]. Several studies have pointed out the use of TQ in the management of Parkinson's disease via reducing lack of climbing ability, oxidative stress, and apoptosis in the brain and also Alzheimer disease by decreasing the expression of β-amyloid [100].

The neuroprotective effects of TQ have been shown by experimental studies, but not yet in clinical trials, and more safety studies should be performed to indicate possible toxic effects of TQ in long-term administration in humans. Conclusively these studies suggests that the neuroprotective effects of TQ are associated with the antioxidant and anti-inflammatory activities. Although experimental studies indicated the beneficial effects of TQ against nervous system problems, better designed clinical trials in humans are needed to confirm these effects.

A significant number of herbal medication and dietary supplements are used for treating patients with neurological or psychiatric complaints. Some of these products may be anticonvulsant and thus of possible benefit in patients with epilepsy. There are more studies that report anti-seizure activity of a nonallopathic preparation with animal models of epilepsy [104]. Beneficial interaction of TQ and sodium valproate in experimental models of epilepsy was demonstrated. This benzoquinone combined with sodium valproate increases its antiepileptic response [105]. Evidence for the pharmacological effect of these components on experimental tests has been provided [106-107]. The pharmacological properties attributed to naturally occurring quinones are thus well established. For example, TQ, the principal active constituent of *N. sativa* seeds, presents anticonvulsant activity in the petit mal epilepsy [108]. Mechanism of action studies has shown that this quinone exerts its anticonvulsant activity through the stimulation of opioid receptors in the central nervous system [109]. These facts led us to verify the pharmacological potential of five structurally related *para*-benzoquinones on the pentylenotetrazol (PTZ)-induced seizures model, and establish the structural characteristics that influence the anticonvulsant activity of TQ. These results indicate TQ as a promising tool, if properly used, in the prevention and treatment of a variety of nerves disorders such as ischemia-reperfusion injury, depression, seizure, Alzheimer and Parkinson's disease. The effect of TQ on different neurological disorders [110] has been listed in Table 3.

Table 3. Effect of TQ on neurological or mental disorders

Neurological or mental Disorders	Model used and intervention (s)	Finding (mechanism)
Alzheimer's disease (AD)	Lipopolysaccharide-induced AD in mice, received TQ (2.5 & 5mg/kg) for 7 days. Aβ-induced neurotoxicity (analyzed by culturing hippocampus and cortical neurons). TQ is administered along with Aβ1−42 for 72 hours	(i) ↓ TBARS & 5-LOX levels (ii) ↑ GSH extent and SOD action (iii) Causes disaggregation of Aβ peptide (iv) prevents declining of neurons (v) Slows degeneration of cognitive ability (i) Reducing Aβ-induced neurotoxicity. (Improved cell viability) by: (ii) Inhibiting mitochondrialmembrane potential depolarization (iii) Hindering reactive oxygen species generation
Parkinson's disease (PD)	1-methyl-4-phenylpyridinium (MPP+) and rotenone-induced neurotoxicity in PD model, cultures were treated with TQ (0.01, 0.1, 1 and, 10 μM) on day 8th for 4 days. Experimental model of early PD induced by 6-hydroxydopamine neurotoxicity, pretreatment of daily TQ (5 & 10 mg/kg) and one additional dose after surgery were used.	(i) Rescued dopaminergic neurons through: (ii) Its antioxidant and anti-inflammatory effects (i) ↓ MDA level (ii) Prevents loss of neurons in substantia nigra (iii) Protects hippocampal & human induced pluripotent stem cell against α-synuclein induced synaptic toxicity
Depression and anxiety	(i) Open field and elevated plus maze models; forced swim test (ii) Locomotor behavior in familiar and new environment in rats, *N. sativa* oil (0.1 mL/day) aqueous seed extract (2 mL/day) orally for 4-6 weeks Stressed and unstressed mice, 10 and 20 mg/kg of TQ for 4 weeks Randomized control trial on healthy human subjects, *N. sativa* capsule (500 mg) daily for 4 weeks.	(i) ↑ in open field activity & struggling time (ii) ↑ 5-HT (iii) ↓ 5HIAA level in the brain (iv) ↑ tryptophan level in plasma & brain (v) ↑ locomotors activity in novel environment (vi) ↑ brain DA level [59, 70, 71] Unstressed mice: at 10 & 20 mg/Kg showed anti-anxiety (i) without altering nitrite levels (ii) ↑ GABA content (only 20mg/Kg). (i) ↓ plasma nitrite level (ii) Reversal of reduced GABA (i) Stabilize disturbed mood (ii) ↓ anxiety (iii) Modulate memory positively
Epilepsy	Pentylenetetrazole-induced seizure, *N. sativa* oil; TQ Double-blinded placebo randomized control trial (refractory epilepsy), TQ as adjunctive therapy for 4 weeks	(i) Prevented seizure occurrence (ii) ↓ Reactive oxygen species generation (iii) Reduced seizure score (iv) Showed additive effects with phenobarbitone (i) Significant reduction of seizure frequency (those who received combination therapy)

↑ = increase, ↓ = decrease. TBARs = Thiobarbituric acid reactive substances, GABA = gamma amino butyric acid, 5-HT= 5 hydroxytryptamine MDA = malondialdehyde, DA = dopamine, 5HIAA = 5 hydroxyindoleacetic acid, GSH = glutathione peroxidase, SOD = superoxide dismutase, TQ = thymoquinone, Aβ = beta amyloid peptides,

3.8. Antidiabetic Effect

Diabetes is a chronic disorder and related to chronic complications for example, problems with the heart functioning, kidney damage, neuropathic pain, and retinopathy. There are so many scientific data available with the traditional or herbal remedies for the diabetes, which makes the development of an alternative therapy for the diabetes and its related complications. TQ has anti-diabetic activity and the most useful and clinically relevance model of human diabetes is streptozotocin (STZ)-induced diabetes in experimental animals. The administration of TQ for one month to streptozotocin-induced diabetic rats displayed a significant reduction of fasting plasma glucose, serum MDA, interleukin-6, and immunoglobulin A, G, and M while substantial increment of endogenous antioxidant enzymes; SOD, Glutathione-S-transferase, and catalase expression were noticed. The histology of pancreas in TQ treated group also revealed an improvement in the pancreatic *β*-cells degeneration, inflammation, and congestion as compared to diabetic control [111]. The marked antidiabetic activity upon three-month supplementation of *N. sativa* (2 g/day) along with oral antidiabetic agent in type 2 DM patients has also been reported. In this study, *N. sativa* received group showed significant reduction of fasting plasma glucose, hemoglobin A1c, and TBARBs, while marked elevation of the total antioxidant capacity, SOD, and glutathione levels were noted [112].

Furthermore, an experimental randomized controlled trial of 99 diabetes patients received the placebo and two treatment groups received oral black seed oil. Administration of 1.5 and 3mL/day of black seed oil for 20 days showed meaningful reduction of glycated hemoglobin A1c and random blood sugar levels [113]. The effect of *N. sativa* seed on the glycemic control of patients with type-2 diabetes (DM-2) was also used as an adjunctive treatment added to their oral hypoglycemic agents. *N. sativa* at a dose of two g/day also influenced substantial reductions in fasting plasma glucose and glycated hemoglobin (HbA1c) without major alteration in body weight [114]. The oil of *N. sativa* at 2 mL/kg also was showed to reduce fasting plasma glucose and intensification of insulin levels in diabetic rats compared to control. Diabetic rats that received NSO exhibited substantial improvements in lipid profile and expressive increment of pancreatic and hepatic antioxidant enzymes also augmented the histological image and glycogen contents other than improvements of average pancreatic islet extent than the diabetic groups [115]. The different doses of *N. sativa* seed (1, 2, and 3 g/day) in patients with DM-2 were also evaluated. A one g/day administration increased high-density lipoprotein cholesterol (HDL-c) levels after 3 months while two and three g/day of *N. sativa* seed significantly decreased serum levels of total cholesterol (TC) and triglyceride (TG) as well as low-density lipoprotein cholesterol (LDL-c) and increased plasma HDL-c [116].

Nutritious supplementation of gestational diabetic mothers through TQ during gestation and lactation ages amended diabetic snags and upheld an effectual T cell resistant reply in their descendants. The uplifting effect of TQ on IL-2 level-cell propagation, and the subsequent liberation of both mingling and thymus homing T cells in the progeny of diabetic mothers heightened the immune retort [117]. Likewise, it designated a lessening in the fraction of abortions, a growth in the number of actual pregnancies and an upgrading of transience amongst new pups innate to diabetic mothers. TQ better divergent hydro peroxide and oxidative stress in pups and these fallouts may be arbitrated by boom in the levels of GST, GSH, CAT, and a reduction in DNA injury [118]. Buildup of progressive glycation end-products in tissues and serum, plays important role in diabetes-associated difficulties. TQ had anti-glycation consequence and abridged diabetic malfunctions due to protein glycation [119]. It controlled the plasma meditations of cholesterol and triglycerides; plasma triglyceride and cholesterol heights reduced in the TQ treated diabetic rats [120]. TQ administered orally at the dose of 10mg/kg ameliorated investigational diabetic neuropathy. TQ showed retrieval of the histopathological changes in sciatic nerves, and myelin break-down diminished evocatively afterward TQ administration [121]. Reactive oxygen species and inflammation play a vital role in the growth of diabetic glitches. STZ-induced diabetes triggered nephropathy, while administration of TQ showed significant improvement in renal morphologic and functional improvement [122]. TQ was operative for b-cell defense in contradiction of injury, through the decreasing inflammatory activity of NO pathway [123]. Intraperitoneal administration of TQ (3 mg/kg) in diabetic rats regularized the raised levels of IL-1b and TNFα [123]. STZ induces a surge in heart and brain NO and MDA, and reduction in endogenous antioxidants, while employment of TQ improved these levels. Serum levels of creatine kinase-MB and brain types was condensed in the diabetic rats, which enhanced with TQ treatment. In diabetes, there was an obvious strengthening in norepinephrine, dopamine and a noticeable decrease in serotonin level; these remained partially up-turned by TQ administration [124].

In reference to modern scholars' devotion to the likely effects of medicinal herbs in diabetic management, a recent meta-analysis of antidiabetic effects of *N. sativa* [125] also exhibited the maintenance of glucose homeostasis and serum lipid profiles in diabetic human subjects [114].

3.9. Nephroprotective Effect

The nephroprotective effects of TQ in various pathogenic conditions have been reported in many studies. TQ found to ameliorate the appearance of renal lesions resulted from various toxic agents partly by weakening reactive oxygen species and inflammation. TQ treatment ameliorated acute renal failure induced by gentamicin in rats by restoring mitochondrial function and salvaging ATP production. It ameliorated degenerative changes induced by gentamicin and improved the renal function markers such as blood urea nitrogen and creatinine. TQ also inhibited lipid peroxidation concomitant to improved antioxidant defense in renal cortex as evidenced by raised GSH level and activities of GSHPx and CAT [126]. TQ prevented mercuric chloride-induced renal damage in rats as evidenced by improved activities of anti-oxidant enzymes and restoration of renal function along with histopathological salvage of renal tissues [34]. Administration of TQ in drinking water (50mg/l) showed to attenuate acute kidney injury induced by cisplatin and improved the therapeutic effects of cisplatin in rodent models. TQ caused reduction in serum urea, creatinine and enhanced polyuria, kidney0 weight and creatinine clearance [78]. It also thwarted vancomycin induced kidney injuries. The antibiotic, vancomycin produced surge in serum blood urea nitrogen, creatinine, and MDA along with reduced activities of SOD and GSHPx in kidney tissues. TQ treatment corrected vancomycin-induced biochemical changes. TQ also improved renal morphology and produced functional improvement in STZ-induced diabetes in rats [121]. TQ (5 mg/kg per day) provided in the drinking water earlier and during ifosfamide treatment found to recover the severity caused by ifosfamide. TQ ameliorated ifosfamide prompted phosphaturia, glycosuria, raised serum creatinine, urea and regulated creatinine clearance rate. It also prevented depletion of GSH from renal tissues and diminished lipid peroxide via antioxidant action [79]. Additionally, TQ also ameliorated doxorubicin (DOX)-induced nephropathy in rats and reduced triglycerides and cholesterol and corrected hyperlipidemia. TQ (10 mg/kg/day) also diminished DOX-induced proteinuria and albuminuria and attenuated total triglycerides, cholesterol and MDA in the kidneys. Besides, non-protein sulfhydryl (NPSH) content and CAT activity in the kidneys of TQ-treated rats obviously raised [31]. TQ treatment abridged kidney damage by regularizing the raised levels of serum urea, creatinine and urinary albumin excretion in DOX induced nephrotoxicity. TQ also subdued lipid peroxidation and enhanced the activities of endogenous anti-oxidant s along with reduced level of renal oxidase NOX-4 [127]. TQ showed effective in hepatorenal dysfunction desirous by kidney I/R. kidney I/R caused in a rise in lipid peroxidation and diminution in GST and SOD in liver and kidney tissues, and TQ treatment produced the reverse of these variations. It abridged spermidine/spermine N-1-acetyltransferase [128]. Treatment with TQ endangered the kidneys after oxidative stress produced by pyelonephritis. In the pyelonephritis, SOD, CAT activity, and lipid peroxidation levels were abnormal and these were refunded to nearly usual and overturned by TQ treatment in the pyelonephritis [129]. TQ supplementation showed protection in cypermethrin induced- sloughing off epithelial cell, shrinkage of glomeruli, and necrosis of renal tubules in kidneys of mice [130]. This suggests that TQ can be promising agent for nephropathies. The effect of TQ on cyclophosphamide-induced hemorrhagic cystitis in mice has been also studied [30]. TQ treatment administered intraperitoneally in the doses of 5, 10, and 20mg/kg before and after the cyclophosphamide injections prevented histological perturbations as evidenced by decreased cellular infiltration, edema, epithelial denudation and hemorrhage in the bladder tissue, reduced oxidative stress, lipid peroxidation and inhibited DNA fragmentation. TQ also induced Nrf2 expression in the bladder tissues of mice concomitant to the restored anti-oxidants in the bladder tissues [131].

3.10. Cardioprotective Effect

Cardiovascular diseases remain one of the leading causes of death worldwide. Several lines of evidence suggest that TQ is a therapeutic option in cardiovascular complications. Ojha and coworkers [132] demonstrated the cardioprotective effects of TQ in isoproterenol-induced myocardial injury in rats. Isoproterenol is known to induce myocardial lesions similar to acute myocardial infarction in rats. The protective effect of TQ against chemical cardiotoxic agents were also investigated in many studies [133]. The therapeutic aspect of TQ (12.5, 25, 50 mg/kg, p.o.) has been investigated against isoproterenol caused myocardial damage. TQ treatment improved the histological modifications, the GSH/GSSG ratio of myocardial, plasma LDH, TBARS, GR and SOD induced by isoproterenol. This study indicated that TQ inhibits cardiotoxicity caused by isoproterenol via enhancing antioxidant content [134]. The cardioprotective effect of TQ (50 mg/L in water, p.o.) against cyclophosphamide has been studied. Cyclophosphamide increased serum cholesterol, TNF-α, TG, CK, LDH, Urea, and creatinine. In addition, cyclophosphamide increased total nitrate/nitrite and TBARS and also decreased CAT, ATP, GPx, and GSH levels of the heart tissue. However, TQ ameliorated the biological alterations caused by cyclophosphamide. Therefore TQ improved CP-induced cardiotoxicity by modulating nitrosative, oxidative stress and ameliorating the mitochondrial function [135]. Doxorubicin (DOX), an antitumor agent, may be disturbing cardiovascular function. One study indicated that TQ (8 mg/kg/d, p.o.) ameliorates cardiotoxic effects of DOX. They indicated that TQ decreases serum LDH and CK levels. They showed that TQ ameliorates cardiotoxicity without changing DOX antitumor effect [136]. TQ (20mg/kg) treatment augmented anti-oxidants and salvaged cardiomyocytes evidenced by restoration of cardiomyocyte injury enzymes and reduction of lipid peroxidation product and proinflammatory cytokines. The findings were further supported by histopathological preservation of the myocardium as evidenced by reduced myonecrosis, edema, and infiltration of inflammatory cells. The findings revealed that TQ exert protective effect on cardiac injury by attenuating oxidative stress, enhancing endogenous anti-oxidants and maintaining structural integrity [132]

3.11. TQ and Apoptosis

Tumor cells tend to elude apoptosis by deregulating genes that perpetuate programmed cell death (apoptosis). Most of the studies are limited to *in vitro* cell experiments, document TQ-mediated apoptosis by regulating multiple targets in the apoptotic machinery. Although evidence for reduced cell viability has been observed in response to TQ treatment in breast, colon, bone, leukemia, larynx, prostate, and Pancreatic cells, the classical hallmark of apoptosis such as chromatin condensation, translocation of phosphatidyl serine across plasma membrane, and DNA fragmentation have been documented in TQ-treated cells. Furthermore, TQ has been shown by studies [137] that it activates the mitochondrial/intrinsic pathway that involves release of cytochrome c from the mitochondria into the cytosol, which in turn binds to the apoptosis protease activation factor-1 (Apaf-1) and leads to the activation of the initiator caspase-9. Activation of caspase-9 has been described following exposure of human myeloblastic leukemia HL-60 cells (8) and PC cells to TQ [138]. Thus, one of the proposed mechanisms of apoptosis induction by TQ involves interference with mitochondrial integrity. There have been no studies reported implicating TQ with the activation of the death receptor/extrinsic pathway of apoptosis in cancer cells. Another molecular entity closed linked to apoptosis is the proapoptotic B-cell non-Hodgkins lymphoma-2 (Bcl-2) family of proteins, which includes Bcl-2-associated x protein (Bax) and Bak, which are activated by TQ pretreatment [139]. In several cell lines, prolonged incubation with TQ showed induction of apoptosis by upregulation of proapoptotic Bax protein along with downregulation of antiapoptotic Bcl-2 proteins, resulting in an enhanced Bax/Bcl-2 ratio. In a recent study reported by Gali-Muhtasib and coworkers [76], checkpoint kinase 1 homolog (CHEK1), a serine/threonine kinase, has been pointed out as one of the targets of TQ, leading to apoptosis in p53+/+ colon cancer cells. On comparing the effect of TQ on p53+/+ as well as p53−/− HCT116 colon cancer cells, p53+/+ cells were found to be more sensitive to TQ in terms of DNA damage and apoptosis induction; it was noted that CHEK1 was nine fold upregulated in p53-null HCT116 cells. The results are in agreement with *in vivo* experimental findings demonstrating that tumors lacking p53 had higher levels of CHEK1, which was associated with poorer apoptosis, advance tumor stages, and worse prognosis. In related context, Alhosin and coworkers [140] studied the effect of TQ on p53 deficient lymphoblastic leukemia Jurkat cells and found TQ treatment produced intracellular ROS promoting a DNA damage-related cell cycle arrest and triggered apoptosis through p73-dependent mitochondrial and cell cycle signaling pathway. This was followed by downregulation of UHRF1, which prevented epigenetic code replication and hindered the cancer signature to be inherited in daughter cancer cells [140]. This highlights a pivotal property of TQ to stimulate cells lacking functional p53 to undergo apoptosis through a p73 component of signaling cascade when p73 is expressed. Overall, the data suggest that TQ inhibited the tumor by up-regulating the tumor suppresser genes such as PTEN and by supressing the Akt/PI3K pathways which is suggestive of its novel therapeutic target and mechanism in cancer.

4. TQ ANALOGS

Attempts have been made to synthesize novel analogs of TQ with superior efficacy for use against tumors. The study done by Effenberger and coworkers [141] recently reported and evaluated conjugates of TQ with various monoterpenes, sesquiterpenes, and the cytotoxic triterpene betulinic acid for improved anticancer activity. They attached esters of various terpene alcohols to C(6) of TQ via spacers of variable length. The resulting derivatives were tested for growth inhibition of cells of the human cancer cell lines HL-60 leukemia, 518A2 melanoma, MDR KB-V1/Vbl cervix carcinoma, and MCF-7/Topo breast adenocarcinoma and nonmalignant foreskin fibroblasts. Some of these analogs were far more efficacious in certain cancer cell lines than the parent drug while being considerably less toxic to nonmalignant human fibroblasts. Among the synthesized derivatives, two conjugates revealed the best results of all test compounds against MDR MCF-7 breast carcinoma cells. Their effects on the mitochondrial membrane potential and the cellular levels of ROS were also evaluated and reported as the basis for their cellular mechanism of action [141]. Studies [142] also recently reported and evaluated the biological effect of TQ analogs against gemcitabine resistant pancreatic cancer cell line (MiaPaCa-2). Out of 27 analogs synthesized, 3 compounds exhibited superior activity than TQ at equimolar concentration (10 μM). These compounds were further evaluated toward improvement of sensitivity to oxaliplatin and gemcitabine in pancreatic cancer cells and again were found superior to parent TQ in causing reduced cell viability and inducing apoptosis [144]. According to the work reported by Ravindran and coworkers [143], TQ encapsulated in biodegradable nanoparticulate formulation (based on poly (lactide-co-glycolide) (PLGA) and the stabilizer polyethylene glycol (PEG)-5000) enhanced antiproliferative, anti-inflammatory, and chemosensitization potential. Further investigations using a panel of markers directed toward understanding its efficacy toward cell proliferation, metastasis, angiogenesis, and chemosensitization revealed TQ nanoparticles (NP) being more active than TQ. TQ-NP were more potent than TQ in suppressing proliferation of colon cancer, breast cancer, prostate cancer, and multiple myeloma cells and was found to be were more potent than TQ in sensitizing leukemic cells to TNF- and paclitaxel-induced apoptosis [144].

5. Need for Clinical Trials of TQ

There is an upsurge in the use of phytomedicines for their therapeutic and preventive benefits. However, the safety issues related with these phytomedicines has also received enormous attention before using in humans. There are plenty of literatures on efficacy but the regulatory toxicology studies are lacking to meet the legal requirements in order to encourage for clinical studies and meet the requirements of their use in health and medical care. It has been revealed that the seeds and oil of the plant comprising TQ demonstration very low grade of deadliness or look barren of toxicities [145]. Numerous studies were approved to evaluate the toxicological chattels of TQ *in vitro* and *in vivo* [145], and lone an imperfect figure of hearsays on the theoretically poisonous possessions of TQ occurs. The animal models of different diseases where TQ has been studied in different dose ranges have been shown a promising preventive and therapeutic agent, endowed with beneficial effects with negligible toxicities [29, 136]. Recently, administration of TQ (20 days) did not encourage death in Balb/c mice or disturb their mean body weight, an actual subtle limit for toxicity in rodent models [146]. TQ treated at 1 mg/kg/day was found fine stood [76] and did not induces any clinically significant change in neurological function, behavioral parameters, biochemical workroom variables or vigorous ciphers. Badary and coworkers [147] demonstrated that TQ (0.03%) supplementary in the drinking water of mice for 3 months did not showed any signs of toxicity, except for a reduction in fasting plasma glucose level. In a study [78], constant distribution of TQ for 30 days by means of tri-calcium phosphate lysine pill (0.02 g of TQ) to adult male rats has shown negligible adverse effects on the dynamic strictures and no sign of toxicity on reproductive organs. It is remarkable that the active amounts of TQ were found safe and TQ even in doses of 90 mg/kg/day upon sub chronic administration to rats found devoid of toxicity. The authors also reported hypo activity and difficulty in respiration as signs of toxicity when TQ administered at high doses (2–3 g/kg) for 24h. It also reduced GSH content in liver, kidneys, and heart tissues. This additional study showed liver and kidney toxicity as established by upsurge in plasma metabolites and enzymes; plasma urea, creatinine, and the enzyme activities of ALT, LDH, and creatine phosphokinase [148].

6. Future Perspectives

This chapter enumerates various properties of TQ and reveal its therapeutic potential against several diseases including diabetes, neuropathic pain, ulcerative colitis, cancer, cardiac, musculoskeletal, and neurodegenerative illnesses including Alzheimer's and Parkinson's disease. TQ being a natural constituent in numerous edible plants makes a dietary component since ancient times and considered safe with time tested evidence. The low toxicity of TQ has gotten it tremendous commercial attention worldwide to be used in foods as well as gaining acceptance for pharmacological research, therapeutic benefits and pharmaceutical development for human application in coming years. Apart from its medicinal use lately, it has also been used in the food industries as an additive or a flavor enhancer or preservative. The available data are suggestive of its use either as nutraceutical or as prophylactic or adjuvant for long-lasting illnesses involving low grade chronic oxidative stress and inflammation. Molecular pharmacology data from several studies depicts that TQ modulates enzymatic, apoptotic, and cell signaling pathway, to elicit its pharmacological effects. The molecular and pharmacological mechanism with its underlying the therapeutic benefits warrants further scientific studies for the pharmaceutical development of TQ. The multifunctional and poly-pharmacological possessions too synergies the action of other drugs and provides a rationale for use in diseases involving multifactorial pathogenesis necessitating one drug many targets approach in therapeutics. Its use with other conventional drugs may also improve the scope of its synergistic blend by enhancing effectiveness and diminishing adverse effects. Although the experimental indications are promising, but there is a need to investigate the translational features of TQ. The precise molecular mechanism by which TQ applies its beneficial effects is still poorly understood. Furthermore, the structure-activity relationships of this pharmacophore need to be investigated in detail for development of druggable congeners. Briefly, the pharmacological properties, favorable pharmacokinetics, relative safety and toxicity, the lipophilicity, high therapeutic index, efficacy and safety profile make TQ a promising candidate for drug development. Notwithstanding the possible excellent advantage of this compound of natural origin, the clinical trials are needed for the translation of the experimental indications into reality in humans.

REFERENCES

[1] Hassanien, M. F. R., Assiri, A. M. A., Alzohairy, A. M. & Oraby, H. F. (2013). Health-promoting value and food applications of black cumin essential oil: an overview, *Journal of Food Science and Technology*, *52*(10), 6136–6142, 2015.

[2] Adib-Hajbaghery, M. & Rafiee, S. (2018). Medicinal plants use by elderly people in Kashan, Iran, *Nursing and Midwifery Studies*, *7* (2), 67–73,

[3] De Luca, V., Salim, V., Atsumi S. M. & Yu, F. (2012). Mining the biodiversity of plants: A revolution in the making, *Science*, *336* (6089), 1658–1661.

[4] Takruri, H. R. H. & Dameh, M. A. F. (1998). Study of the nutritional value of black cumin seeds (Nigella sativa L) *Journal of the Science of Food and Agriculture*, *76* (3), 404–410.

[5] Salem, M. L. (2005). Immunomodulatory and therapeutic properties of the Nigella sativa L. seed, *International Journal of Immunopharmacology*, *5*, 1749–1770, PMID: 16275613.

[6] Ramadan, M. F. (2007). Nutritional value, functional properties and nutraceutical applications of black cumin (Nigella sativa L.): an overview, *International Journal of Food Science & Technology*, *42*(10), 1208–1218.

[7] Gali-Muhtasib, H., El-Najjar, N. & Schneider-Stock, R. (2006). The medicinal potential of black seed (Nigella sativa) and its components, *Advances in Phytomedicine*, *2*, 133–153.

[8] El-Dakhakhny, M. (1963). Studies on the chemical constitution of Egyptian *N. sativa* L. seeds, *Planta Medica.*, *11*, 465–470.

[9] Khader, M. & Eckl, P. M. (2014). Thymoquinone: an emerging natural drug with wide range of medical applications. *Iranian Journal of Basic Medical Sciences*, *17* (12), 950-957.

[10] Hirobe, C., Qiao, Z. S., Takeya, K. & Itokawa, H. (1998). Cytotoxic principles from Majorana syriaca. *Journal of Medicinal Chemistry*, *52*, 74–77.

[11] Economakis, C., Skaltsa, H., Demetzos, C., Sokovic, M. & Thanos, C. A. (2002). Effect of phosphorus concentration of the nutrient solution on the volatile constituents of leaves and bracts of *Origanum dictamnus*. *Journal of Agricultural Food and Chemistry*, *50*, 6276–6280.

[12] Ipek, E., Zeytinoglu, H., Okay, S., Tuylu, B. A., Kurkcuoglu, M. & Baser, K. H. C. (2005). Genotoxicity and antigenotoxicity of Origanum oil and carvacrol evaluated by Ames *Salmonella/* microsomal test. *Food Chemistry.*, *93*, 551–556.

[13] Lukas, B., Schmiderer, C., Franz, C. & Novak, J. (2009). Composition of essential oil compounds from different Syrian populations of *Origanum syriacum* L. (*Lamiaceae*). *Journal of Agricultural Food and Chemistry*, *57*, 1362–1365.

[14] Burits, M. & Bucar, F. (2000) Antioxidant activity of *Nigella sativa* essential oil. *Phytotherapy Research*, *14*, 323–328.

[15] Singh, S., Das, S. S., Singh, G., Schuff, C., Lampasona, M. P. de. & Catalan, C. A. N. (2014). Composition, *in vitro* antioxidant and antimicrobial activities of essentialoil and oleoresins obtained from

black cumin seeds (*Nigella sativa* L.). *Biomed Research International*, Article ID 918209, 10pages.

[16] Solati Z., Baharin, B. S. & Bagheri, H. (2014). Antioxidant property, thymoquinone content and chemical characteristics of different extracts from *Nigella sativa* L. seeds *Journal of the American Oil Chemists' Society*, *91* (2), 295–300.

[17] Herlina, Aziz S. A., Kurniawati, A. & Faridah, D. N. (2017). Changes of thymoquinone, thymol and malondialdehyde content of black cumin (*Nigella sativa* L.) in response to Indonesia tropical. *Journal of Biosciences*, *24* (3), 156–161.

[18] Salea, R., Widjojokusumo, E., Hartanti, A. W., Veriansyah, B. & Tjandrawinata, R. R. (2013). Supercritical fluid carbon dioxide extraction of *Nigella sativa* (black cumin) seeds using taguchi method and full factorial design, *Biochemical Compounds*, *1*, 1-7.

[19] Matthaus, B. & Ozcan, M. M. (2011). Fatty acids, tocopherol, and sterol contents of some nigella species seed oil *Czech Journal of Food Sciences*, *29* (2), 145–150.

[20] Mehta, B. K., Verma, M. & Gupta, M. (2008). Novel lipid constituents identified in seeds of *Nigella sativa* (Linn). *Journal of the Brazilian Chemical Society*, *19*(3), 458–462.

[21] San Mauro-Mart´ın, I., Blumenfeld-Olivares, J. A., Garicano- Vilar, E., Cuadrado, M. A., Ciudad-Caba˜nas, M. J. & Collado-Yurrita, L. (2018). Differences in the effect of plant sterols on lipid metabolism in men and women. *Topics in Clinical Nutrition*, *33* (1), 31–40.

[22] Cheikh-Rouhou, S., Besbes, S., Hentati, B., Blecker, C., Deroanne, C. & Attia, H. (2007). *Nigella sativa* L.: Chemical composition and physicochemical characteristics of lipid fraction. *Food Chemistry*, *101* (2), 673–681.

[23] Zaunschirm, M., Pignitter, M., Kienesberger J., Hernler, N., Riegger, C., Eggersdorfer, M. & Somoza, V. (2018). Contribution of the ratio of tocopherol homologs to the oxidative stability of commercial vegetable oils," *Molecules*, *23* (1), 206.

[24] Morikawa, T., Xu, F., Kashima, Y., Matsuda, H., Ninomiya, K. & Yoshikawa, M. (2004). Novel dolabellane-type diterpene alkaloids

with lipid metabolism promoting activities from the seeds of *Nigella sativa*. *Organic Letters*, *6*(6), 869–872.

[25] Morikawa, T., Xu F., Ninomiya K., Matsuda, H. & Yoshikawa, M. (2004). Nigellamines A3, A4, A5, and C, new dolabellane type diterpene alkaloids, with lipid metabolism-promoting activities from the Egyptian medicinal food black cumin," *Chemical & Pharmaceutical Bulletin*, *52* (4), 494–497.

[26] Ali, Z., Ferreira, D., Carvalho, P., Avery, M. A. & Khan, I. A. (2008). Nigellidine-4-o-sulfite, the first sulfated indazole-type alkaloid from the seeds of Nigella sativa," *Journal of Natural Products*, *71* (6), 1111-1112.

[27] Alkharfy, K. M., Ahmad, A., Khan, R. M. & Al-Asmari, M. (2013). High performance liquid chromatography of thymoquinone in rabbit plasma and its application to pharmacokinetics. *Journal of Liquid Chromatography and Related Technologies*, *36*, 2242–2250.

[28] Salmani, J. M., Asghar, S., Lv, H. & Zhou, J. (2014). Aqueous solubility and degradation kinetics of the phytochemical anticancer thymoquinone; probing the effectsof solvents, pH and light, *Molecules*, *19*, 5925–5939, PMID: 24815311.

[29] Mansour, M., Ginawi, O., El-Hadiyah, T., El-Khatib, A., Al-Shabanah, O. & Al-Sawaf, H. (2001). Effects of volatile oil constituents of *Nigella sativa* on carbon tetrachloride-induced hepatotoxicity in mice: evidence for antioxidant effects of thymoquinone. *Research Communications in Molecular Pathology and Pharmacology*, *110*, 239–252.

[30] Goyal, S. N., Prajapati, C. P., Gore, P. R., Patil, C. R., Mahajan, U. B., Sharma, C., Talla, S. P. & Ojha, S. K. (2017). Therapeutic Potential and Pharmaceutical Development of Thymoquinone: A Multitargeted Molecule of Natural Origin. *Frontiers in Pharmacology*, *8*, 656.

[31] Badary, O. A., Abdel-Naim, A. B., Abdel-Wahab, M. H. & Hamada, F. M. (2000). The influence of thymoquinone on doxorubicin induced hyperlipidemic nephropathy in rats. *Toxicology*, *143*, 219–226

[32] Khan, M. A., Anwar, S., Aljarbou, A. N., Al-Orainy, M., Aldebasi, Y. H., Islam, S., et al. (2014). Protective effect of thymoquinone on glucose or methylglyoxal induced glycation of superoxide dismutase. *International Journal of Biological Macromolecules*, *65*, 16–20.

[33] Khalife, K. & Lupidi, G. (2007). Nonenzymatic reduction of thymoquinone in physiological conditions. *Free Radicals Research*, *41*, 153–161.

[34] Fouda, A. M. M., Daba, M. H. Y., Dahab, G. M. & Sharaf El-Din, O. A. (2008). Thymoquinone ameliorates renal oxidative damage and proliferative response induced by mercuric chloride in rats. *Basic & Clinical and Pharmacology Toxicology*, *10*, 109–118.

[35] Harzallah, H. J., Grayaa. R., Kharoubi, W., Maaloul, A., Hammami, M. & Mahjoub, T. (2012). Thymoquinone, the *Nigella sativa* bioactive compound, prevents circulatory oxidative stress caused by 1, 2- dimethylhydrazine in erythrocyte during colon post initiation carcinogenesis. *Oxidative Medicine and Cellular Longevity*, *854065.*

[36] Alenzi, F. Q., El-Bolkiny, Y. & Salem, M. L. (2010). Protective effects of *Nigella sativa* oil and thymoquinone against toxicity induced by the anti-cancer drug cyclophosphamide. *Brazilian Journal of Biomedical Sciences*, *67*, 20–28.

[37] Alhebshi, A. H., Gotoh, M. & Suzuki, I. (2013). Thymoquinone protects cultured rat primary neurons against amyloid beta-induced neurotoxicity. *Biochemical and Biophysics Research Communications*, *433*, 362–367.

[38] El Gazzar, M., El Mezayen, R., Marecki, J. C., Nicolls, M. R., Canastar, A. & Dre-skin, S. C. (2006). Anti-inflammatory effect of thymoquinone in a mouse model of allergic lung inflammation. *International Journal of Immunopharmacology*, *6*, 1135–1142.

[39] Kundu, J., Kim, D. H., Kundu, J. K. & Chun, K. S. (2014). Thymoquinone induces hemeoxygenase-1 expression in HaCaT cells via Nrf2/ARE activation: Akt and AMPK α as upstream targets, *Food and Chemical Toxicology*, *65*, 18–26.

[40] Bai, T., Yang, Y., Wu, Y. L., Jiang, S., Lee, J. J., Lian, L. H. & Nan, J. X. (2014). Thymoquinone alleviates thioacetamide induced hepatic

fibrosis and inflammation by activating LKB1-AMPK signaling pathway in mice. *International Journal of Immunopharmacology*, *19*, 351–357.

[41] Kundu, J. K., Liu, L., Shin, J. W. & Surh, Y. J. (2013). Thymoquinone inhibits phorbol ester-induced activation of NF kappa B and expression of COX-2, and induces expression of cytoprotective enzymes in mouse skin *in vivo*. *Biochemical and Biophysical Research Communications*, *438*, 721–727.

[42] Al Wafai, R. J. (2013). *Nigella sativa* and thymoquinone suppress cyclooxygenase-2 and oxidative stress in pancreatic tissue of streptozotocin-induced diabetic rats. *Pancreas*, *42*, 841–849.

[43] Kanter, M., Demir, H., Karakaya, C. & Ozbek, H. (2005). Gastroprotective activity of *Nigella sativa* L oil and its constituent, thymoquinone against acute alcohol-induced gastric mucosal injury in rats. *World Journal of Gastroenterology*, *11*, 6662–6666.

[44] Umar, S., Zargan, J., Umar, K., Ahmad S., Katiyar, C. K. & Khan, H. A. (2012). Modulation of the oxidative stress and inflammatory cytokine response by thymoquinone in the collagen induced arthritis in Wistar rats. *Chemico- Biological Interactions*, *197*, 40–46.

[45] Nemmar, A., Al-Salam, S., Zia, S., Marzouqi, F., Al-Dhaheri, A., Subramaniyan, D., Yasin, J., Ali, B. H. & Kazzam, E. E. (2011). Contrasting actions of diesel exhaust particles on the pulmonary and cardiovascular systems and the effects of thymoquinone. *Brazilian Journal of Pharmacology.*, *164*, 1871–1882.

[46] El Gazzer, M., El Mezayen, R., Nicolls, M. R., Marecki, J. C. & Dreskin, S. C. (2006). Down-regulation of leukotriene biosynthesis by thymoquinone attenuates airway inflammation in a mouse model of allergic asthma. *Biochimica Biophysica Acta*, *1760*, 1088–1095.

[47] Ammar El, S. M., Gameil, N M., Shawky, N. M. & Nader, M. A. (2011). Comparative evaluation of anti-inflammatory properties of thymoquinone and curcumin using an asthmatic murine model. *International Journal of Immunopharmacology*, *11*, 2232–2236.

[48] Boskabady, M. H., Vahedi, N., Amery, S. & Khakzad, M. R. (2011). The effect of *Nigella sativa* alone, and in combination with

dexamethasone, on tracheal muscle responsiveness and lung inflammation in sulfur mustard exposed guinea pigs. *Journal of Ethnopharmacology*, *137*, 1028–1034.

[49] Keyhanmanesh, R., Boskabady, M. H., Ebrahimi Saadatloo, M. A. & Khamnei, S. (2007). The contribution of water and lipid soluble substances in the relaxant effects of *Nigella sativa* extract on guinea pig tracheal smooth muscle (*in vitro*). *Iranian Journal of Basic Medical Sciences*, *10*, 154–161.

[50] Vaillancourt, F., Silva, P. Shi Q., Fahmi, H., Fernandes, J. C. & Benderdour, M. (2011). Elucidation of molecular mechanisms underlying the protective effects of thymoquinone against rheumatoid arthritis. *Journal of Cellular Biochemistry*, *112*, 107–117.

[51] Li, F., Rajendran, P. & Sethi, G. (2010). Thymoquinone inhibits proliferation, induces apoptosis and chemosensitizes human multiple myeloma cells through sup-ression of signal transducer and activator of transcription 3 activation pathway. *Brazilian Journal of Pharmacology*, *161*, 541–554.

[52] Mohamed, A., Afridi, D. M., Garani, O. & Tucci, M. (2005). Thymoquinone inhibits the activation of NF-kappa B in the brain and spinal cord of experimental autoimmune encephalomyelitis. *Biomedical Sciences Instrumentation.*, *41*, 388–393,

[53] Mahgoub, A. A. (2003). Thymoquinone protects against experimental colitis in rats. *Toxicology Letters*, *143*, 133–143.

[54] El-Mahmoudy, A., Matsuyama, H., Borgan, M. A., Shimizu, Y., El-Sayed, M. G., Minamoto, N. & Takewaki, T. (2002). Thymoquinone suppresses expression of inducible nitric oxide synthase in rat macrophages, *International journal of Immunopharmacology*, *2*, 1603-1611.

[55] Oguz, S., Kanter M., Erboga, M. & Erenoglu, C. (2012) Protective effects of thymoquinone against cholestatic oxidative stress and hepatic damage after biliary obstruction in rats. *Journal of Molecular Histology*, *43*, 151–159.

[56] Marsik, P., Kokoska, L., Landa, P., Nepovim, A., Soudek, P. & Vanek, T. (2005). *In vitro* inhibitory effects of thymol and quinones

of Nigella sativa seeds oncyclooxygenase-1- and -2-catalyzed prostaglandin E2 biosyntheses. *Planta Medica*, *71*, 739–742.

[57] El-Mahmoudy, A., Shimizu, Y., Shiina, T., Matsuyama, H., El-Sayed, M. & Takewaki, T. (2005). Successful abrogation by thymoquinone against induction of diabetes mellitus with streptozotocin via nitric oxide inhibitory mechanism. *International journal of Immunopharmacology*, *5*, 195–207.

[58] Haq, A., Lobo, P. I., Al-Tufail, M., Rama, N. R. & Al-Sedairy, S. T. (1999). Immunomodulatory effect of *Nigella sativa* proteins fractionated by ion exchange chromatography. *International journal of Immunopharmacology*, *21*, 283–295.

[59] Xuan, N. T., Shumilina, E., Qadri, S. M., Götz, F. & Lang, F. (2010). Effect of thymoquinone on mouse dendritic cells. *Cellular Physiology and Biochemistry*, *25*, 307–314.

[60] Yang, W., Bhandaru, M., Pasham, V., Bobbala, D., Zelenak C., Jilani, K., Rotte, A. & Lang, H. (2012). Effect of thymoquinone on cytosolic pH and Na+/H+ exchanger activity in mouse dendritic cells. *Cellular Physiology and Biochemistry*, *29*, 21–30,

[61] Reddy, L., Odhav, B. & Bhoola, K. D. (2003). Natural products for cancer prevention: a global perspective. *Pharmacology Therapy*, *99*, 1–13.

[62] Shoieb, A. M., Elgayyar, M., Dudrick, P. S., Bell, J. L. & Tithof, P. K. (2003). *In vitro* inhibition of growth and induction of apoptosis in cancer cell lines by thymoquinone, *International Journal of Oncology*, *22*, 107–113.

[63] Khader, M., Eckl, P. M. & Bresgen N. Effects of aqueous extracts of medicinal plants on MNNG-treated rat hepatocytes in primary cultures. *Journal of Ethnopharamacology.*, *112*, 199–202.

[64] El-Mahdy, M. A., Zhu, Q., Wang, Q. E., Wani, G. & Wani, A. A. (2005). Thymoquinone induces apoptosis through activation of caspase-8 and mitochondrial events in p53-null myeloblastic leukemia HL-60 cells. *International Journal of Cancer.*, *117*, 409–417.

[65] Ulasli, S. S., Celik, S., Gunay, E., Ozdemir, M., Hazman, O., Ozyurek, A., Koyuncu, T. & Unlu, M. Anti-cancer effects of thymoquinone, caffeic acid phenethyl ester and resveratrol on A549 non-small cell lung cancer cells exposed to benzo(a)pyrene. *Asian Pacific Journal of Cancer Prevention.*, *14*, 6159–6164.

[66] Woo, C. C., Loo, S. Y., Gee, V., Yap, C. W., Sethi, G. & Kumar, A. P. & Tan, K. H. (2011). Anti-cancer activity of thymoquinone in breast cancer cells: possible involvement of PPAR-gamma pathway. *Biochemical Pharmacology*, *82*, 464–475.

[67] Woo, C. C., Hsu, A., Kumar, A. P., Sethi, G. & Tan, K. H. (2013). Thymoquinone inhibits tumor growth and induces apoptosis in a breast cancer xenograft mouse model: the role of p38 MAPK and ROS. *PLoS One.*, *8*, e75356.

[68] Xu, D., Ma, Y., Zhao, B., Li, S., Zhang, Y., Pan, S., Wu, Y., Wang, J., Wang, D., Pan, H., Liu, L. & Jiang, H. (2014). Thymoquinone induces G2/M arrest, inactivates PI3K/Akt and nuclear factor-kappaB pathways in human cholangiocarcinomas both *in vitro* and *in vivo*. *Oncology Reports*, *31*, 2063–2070.

[69] Ahmad, I., Muneer, K. M., Tamimi, I. A., Chang, M. E., Ata, M. O. & Yusuf, N. (2013). Thymoquinone suppresses metastasis of melanoma cells by inhibition of NLRP3 inflammasome. *Toxicology and Applied Pharmacology*, *270*, 70–76.

[70] Acharya, B. R., Chatterjee, A., Ganguli, A., Bhattacharya, S. & Chakrabarti, G. (2014). Thymoquinone inhibits microtubule polymerization by tubulin binding and causes mitotic arrest following apoptosis in A549 cells. *Biochimie*, *97*, 78–91.

[71] Alhosin, M., Ibrahim, A., Boukhari, A., Sharif, T., Gies, J. P., Augr, C. & Shini-Kerth, V. B. (2012). Anti-neoplastic agent thymoquinone induces degradation of alpha and beta tubulin proteins in human cancer cells without affecting their level in normal human fibroblasts. *Investigational New Drugs.*, *30*, 1813–1819.

[72] Ashley, R. E. & Osheroff, N. (2014). Natural products as topoisomerase II poisons: effects of thymoquinone on DNA cleavage

mediated by human topoisomerase II alpha. *Chemical Research in Toxicology*, *27*, 787–793.

[73] Racoma, I. O., Meisen, W. H., Wang, Q. E., Kaur, B. & Wani, A. A. (2013). Thymoquinone inhibits autophagy and induces cathepsin-mediated, caspase-independent cell death in glioblastoma cells. *PLoS One.*, *8*, e72882.

[74] Paramasivam, A., Sambantham, S, Shabnam, J., Raghunandha kumar, S., Anandan, B., Rajiv, R., Vijayashree Priyadharsini, J. & Jayaraman, G. (2012). Anti-cancer effects of thymoquinone in mouse neuroblastoma (Neuro 2-a) cells through caspase-3 activation with down-regulation of XIAP. *Toxicology Letters*, *213*, 151–159.

[75] Helal, G. K. (2010). Thymoquinone supplementation ameliorates acute endotoxemia-induced liver dysfunction in rats. *Pakistan Journal of Pharmaceutical Sciences*, *23*, 131–137.

[76] Gali-Muhtasib, H., Ocker, M., Kuester, D., Krueger, S., El-Hajj, Z., Diestel, A., Evert, M., El-Najjar, N., Peters, B., Jurjus, A., Roessner, A. & Schneider-Stock, R. Thymoquinone reduces mouse colon tumor cell invasion and inhibits tumor growth in murine colon cancer models. *Journal of Cellular and Molecular Medicine*, *12*, 330–342.

[77] Badary, O. A. & Gamal El-Din, A. M. (2001). Inhibitory effects of thymoquinone against 20- methylcholanthrene-induced fibrosarcoma tumorigenesis. *Cancer Detection and Prevention*, *25*, 362–368.

[78] Badary, O. A., Nagi, M. N., Al-Shabanah, O. A., Al-Sawaf, H. A., Al-Sohaibani, M. O. & Al-Bekairi, A. M. (1997). Thymoquinone ameliorates the nephrotoxicity induced by cisplatin in rodents and potentiates its antitumor activity. *Canadian Journal of Physiology and Pharmacology*, *75*, 1356–1361.

[79] Badary, O. A. (1999). Thymoquinone attenuates ifosfamide-induced Fanconi syndrome in rats and enhances its antitumor activity in mice. *Journal of Ethnopharmacology*, *67*, 135–142.

[80] Kaseb, A. O., Chinnakannu, K., Chen, D., Sivanandam, A., Tejwari, S., Menon, M., Dou, Q. P. & Reddy, G. P. (2007). Androgen receptor and E2F-1 targeted thymoquinone therapy for hormone-refractory prostate cancer. *Cancer Research*, *67*, 7782– 7788.

[81] Chehl, N., Chipitsyna G, Gong, Q., Yeo, C. J. & Arafat, H. A. (2009). Anti-inflammatory effects of the *Nigella sativa* seed extract, thymoquinone, in pancreatic cancer cells. *HPB (Oxford)*, *11*, 373–381.

[82] Siveen, K. S., Mustafa, N., Li, F., Kannaiyan, R., Ahn, K. S., Kumar, A. P., Ching, W. & Sethi, G. (2014). Thymoquinone overcomes chemoresistance and enhances the anti-cancer effects of bortezomib through abrogation of NF-kappa B regulated gene products in multiple myeloma xenograft mouse model. *Oncotarget*, *5*, 634–648.

[83] Ashour, A. E., Abd-Allah, A. R., Korashy. H. M., Attia, S. M., Alzahrani, A. Z., Saquib, Q., Bakheet, S. A., Abdel-Hamied, H. E., Jamal, S. & Rishi, A. K. (2014). Thymoquinone suppression of the human hepatocellular carcinoma cell growth involves inhibition of IL-8 expression, elevated levels of TRAIL receptors, oxidative stress and apoptosis. *Molecular and Cellular Biochemistry*, *389*, 85–98.

[84] Kouidhi, B., Zmantar, T., Jrah, H., Souiden, Y., Chaieb, K., Mahdouani, K. & Bakhrouf, A. (2011). Antibacterial and resistance-modifying activities of thymoquinone against oral pathogens. *Annals of Clinical Microbiology and Antimicrobials*, *10*, 29.

[85] Harzallah, H. J., Kouidhi, B., Flamini, G., Bakhrouf, A. & Mahjoub, T. (2011). Chemical composition, antimicrobial potential against cariogenic bacteria and of Tunisian *Nigella sativa* essential oil and thymoquinone. *Food Chemistry*, *129*, 1469–1474.

[86] Chaieb, K., Kouidhi, B., Jrah, H., Mahdouani, K. & Bakhrouf, A. (2011). Antibacterial activity of Thymoquinone, an active principle of Nigella sativa and its potency to prevent bacterial biofilm formation. *BMC Complementary and Alternative Medicine*, *11*, 29.

[87] Halawani, E. (2009). Antibacterial activity of thymoquinone and thymohydroquinone of *Nigella sativa* L. and their interaction with some antibiotics. *Advances in Biological Research*, *3*, 148–152.

[88] Aljabre, S. H. M., Randhawa, M. A., Akhtar, N., Alakloby, O. M., Alqurashi, A. M. & Aldossary, A. (2005). Antidermatophyte activity of ether extract of *Nigella sativa* and its active principle, thymoquinone. *Journal of Ethnopharmacology*, *101*, 116–119.

[89] Cingi, C., Eskiizmir, G., Buruko˘glu, D., Erdo˘gmu¸s, N., Ural, A. & Ünlü, H. (2011). The histopathological effect of thymoquinone on experimentally induced rhinosinusitis in rats. *American Journal of Rhinology & Allergy*, *25*, e268–e272.

[90] Sheikh, B. Y., Taha, M. M. E., Koko, W. S. & Abdelwahab, S. I. (2015). Antimicrobial effects of thymoquinone on *Entamoeba histolytica* and *Giardia lamblia. Pharmacology Journal*, *8*, 168–170.

[91] Umar, S., Shah, M. A., Munir, M. T., Yaqoob, M., Fiaz, M., Anjum, S., Kaboudi, K., Bouzouaia, M., Younus, M., Nisa, Q., Iqbal, M. & Umar, W. (2016). Synergistic effects of thymoquinone and curcumin on immune response and anti-viral activity against avian influenza virus (H9N2) in turkeys. *Poultry Science*, *95*, 1513–1520.

[92] Forouzanfar, F., Fazly Bazzaz, B. S. & Hosseinzadeh, H. (2014). Black cumin (Nigella sativa) and its constituent (thymoquinone): A review on antimicrobial effects. *Iran Journal of Basic Medical Sciences*, *17*, 929-938.

[93] El-Dakhakhny, M., Barakat, M., El-Halim, M. A. & Aly, S. M. (2000). Effects of *Nigella sativa* oil on gastric secretion and ethanol induced ulcer in rats. *Journal of Ethnopharmacology*, *72* (1-2), 299-304.

[94] Kalus, U., Pruss, A., Bystron, J., Jurecka, M., Smekalova, A., Lichius, J. J. & Kiesewetter, H. (2003). Effect of *Nigella sativa* (black seed) on subjective feeling in patients with allergic diseases. *Phytotherapy Research*, *17*(10), 1209-1214.

[95] Al-Majed, A. A., Daba, M. H., Asiri, Y. A., Al-Shabanah, O. A., Mostafa, A. A. & El-Kashef, H. A. (2001). Thymoquinone-induced relaxation of guinea-pig isolated trachea. *Research Communications in. Molecular Pathology and Pharmacology*, *110*(5-6), 333-345.

[96] Tahereh Farkhondeh, T., Samarghandian, S., Shahri, A. M. P. & Samini, F. (2018). The neurprotective effects of Thymoquinone: A review. *Dose-Response*, *16*(2), 1559325818761455.

[97] Constantinescu, C. S., Farooqi, N., O'brien, K. & Gran, B. (2011). Experimental autoimmune encephalomyelitis (EAE) as a model for

multiple sclerosis (MS). *Brazilian Journal of Pharmacology*, *164*(4), 1079–1106.

[98] Radad, K., Moldzio, R., Taha, M. & Rausch, W. D. (2009). Thymoquinone protects dopaminergic neurons against MPP+ and rotenone. *Phytotherapy Research*, *23*(5), 696–700.

[99] Liu, J., Wang, A., Li, L., Huang, Y., Xue, P. & Hao, A. (2010). Oxidative stress mediates hippocampal neuron death in rats after lithium–pilocarpine-induced status epilepticus. *Seizure*, *19*(3), 165–172.

[100] Ismail, N., Ismail, M., Mazlan, M., Latiff, L. A., Imam, M. U., Iqbal, S., Azmi, N. H., Ghafar, S. A. & Chan, K. W. (2013). Thymoquinone prevents β-amyloid neurotoxicity in primary cultured cerebellar granule neurons. *Cell Mol Neurobiol.*, *33*(8), 1159–1169

[101] Espinosa-Diez, C., Miguel, V., Mennerich, D., Kietzmann, T., Sánchez-Pérez, P., Cadenas, S. & Lamas, S. (2015). Antioxidant responses and cellular adjustments to oxidative stress. *Redox Biology*, *6*, 183–197.

[102] Samarghandian, S., Borji, A., Afshari, R., Delkhosh, M. B. & Gholami, A. (2013). The effect of lead acetate on oxidative stress and antioxidant status in rat bronchoalveolar lavage fluid and lung tissue. *Toxicology Mechanism Methods*, *23*(6), 432–436.

[103] Darakhshan, S., Pour, A. B., Colagar A. H. & Sisakhtnezhad, S. (2010). Thymoquinone and its therapeutic potentials. *Pharmacological Research*, 95–96, 138–158.

[104] Tyagi, A. & Delanty, N. (2003). Herbal remedies, dietary supplements, and seizures. *Epilepsia*, *44*, 228-235.

[105] Raza, M., Alghasham, A. A., Alorainy, M. S. & El-Hadiyah, T. M. (2006). Beneficial interaction of thymoquinone and sodium valproate in experimental models of epilepsy: reduction in hepatotoxicity of valproate. *Scientia Pharmaceutica*, *74*, 159-173.

[106] De Almeida, R. N., De Sousa, D. P., Nóbrega, F. F. F., Claudino, F. S., Araújo, D. A. M., Leite, J. R. & Mattei, R. (2008). Anticonvulsant effect of a natural compound α,β-epoxy-carvone and its action on the nerve excitability. *Neuroscience Letters*, *443*, 51-55.

[107] De Sousa, D. P., Quintans, J. L. & Almeida, R. N. (2007). Evolution of the anticonvulsant activity of alfa-terpineol. *Pharmaceutical Biology*, *45*, 69-70.

[108] Hosseinzadeh, H. & Parvardeh, S. (2004). Anticonvulsant effects of thymoquinone, the major constituent of *Nigella sativa* seeds, in mice. *Phytomedicine 11*, 56-64.

[109] Hosseinzadeh, H., Parvardeh, S., Nassiri-Asl, M. & Mansouri, M. (2005). Intracerebroventricular administration of thymoquinone, the major constituent of *Nigella sativa* seeds, suppresses epileptic seizures in rats. *Medical Science Monitor*, *11*, BR106-110.

[110] Yimer, E. M., Beshir Tuem, K., Karim, A., Najeeb Ur-Rehman, N. & Anwar, F. (2019). *Nigella sativa* L. (Black Cumin): A Promising Natural Remedy for Wide Range of Illnesses. *Evidence-Based Complementary and Alternative Medicine*, Article ID 1528635, 16 pages.

[111] El Rabey, H. A., Al-Seeni, M. N. & Bakhashwain, A. S. (2017). The antidiabetic activity of *Nigella sativa* and propolis on streptozotocin-induced diabetes and diabetic nephropathy in male rats. *Evidence-Based Complementary and Alternative Medicine*, Article ID 5439645, 14 pages.

[112] Kaatabi H., Bamosa A. O., Badar, A., Al-Khadra, A., Al Elq, A., Abou-Hozaifa, B., Lebda, F. & Al-Almaie, S. (2015). *Nigella sativa* improves glycemic control and ameliorates oxidative stress in patients with type 2 diabetes mellitus: Placebo controlled participant blinded clinical trial. *PLoS ONE*, vol. *10* (2).

[113] Rachman, P. N. R. & Akrom and Darmawan E. (2017). The efficacy of black cumin seed (*Nigella sativa*) oil and hypoglycemic drug combination to reduce HbA1c level in patients with metabolic syndrome risk. in *Proceedings of the International Pharmacy Conference*, vol. *259*, Indonesia, 2017.

[114] Bamosa, A., Kaatabi, H., Badar, A., Al-Khadra, A., Al Elq, A., Abou-Hozaifa, B., Lebda, F. & Al-Almaie, S. (2015). *Nigella sativa*: A potential natural protective agent against cardiac dysfunction in

patients with type 2 diabetes mellitus. *Journal of Family and Community Medicine*, *22* (2), 88–95.

[115] Abdelrazek, H. M. A., Kilany, O. E., Muhammad, M. A. A., Tag H. M. & Abdelazim, A. M. (2018). Black seed and thymoquinone improved insulin secretion, hepatic glycogen storage, and oxidative stress in streptozotocin-induced diabetic male Wistar rats," *Oxidative Medicine and Cellular Longevity*, Article ID 8104165, 10 pages. 2018.

[116] Kaatabi, H., Bamosa, A. O., Lebda, F. M., Al ELq, A. H. & Al-Sultan Al. (2012). Favorable impact of Nigella sativa seeds on lipid profile in type 2 diabetic patients. *Journal of Family and Community Medicine*, *19* (3), 155–160.

[117] Badr, G., Alwasel, S., Ebaid, H., Mohany, M. & Alhazza, I. (2011). Perinatal supplementation with thymoquinone improves diabetic complications and T cell immune responses in rat offspring. *Cellular Immunology*, *267*, 133–140.

[118] Kapoor, S. (2009). Emerging clinical and therapeutic applications of Nigella sativa in gastroenterology. *World Journal of Gastroenterology*, *15*, 2170–2171.

[119] Khan, M. A., Anwar, S., Sadaf, A. & Younus, H. (2014). A structural study on the protection of glycation of superoxide dismutase by thymoquinone. *International Journal of Biological Macromolecules*, *69*, 476–481.

[120] Al-Naqeep, G., Ismail, M. & Yazan, L. S. (2009). Effects of thymoquinone rich fraction and thymoquinone on plasma lipoprotein levels and hepatic low density lipoprotein receptor and 3-hydroxy-3-methylglutaryl coenzyme A reductase genes expression. *Journal of Functional Foods*, *1*, 298–303.

[121] Kanter, M. (2009). Protective effects of thymoquinone on streptozotocin induced diabetic nephropathy. *Journal of Molecular Histology*, *40*, 107–115.

[122] Pye, C., Elsherbiny, N. M., Ibrahim, A. S., Liou, G. I., Chadli, A., Al- Shabrawey, M. & Elmarakby, A. A. (2014). Adenosine kinase inhibition protects the kidney against streptozotocin-induced

diabetes through anti-inflammatory and antioxidant mechanisms. *Pharmacology Research*, *85*, 45–54.

[123] El-Mahmoudy, A., Shimizu, Y., Shiina, T., Matsuyama, H., El-Sayed, M. & Takewaki, T. (2005). Successful abrogation by thymoquinone against induction of diabetes mellitus with streptozotocin via nitric oxide inhibitory mechanism. *International Immunopharmacology*, *5*, 195–207.

[124] Hamdy, N. M. & Taha, R. A. (2009). Effects of *Nigella sativa* oil and thymoquinone on oxidative stress and neuropathy in streptozotocin-induced diabetic rats. *Pharmacology*, *84*, 127–134.

[125] Daryabeygi-Khotbehsara, R., Golzarand, M., Ghaffari, M. P. & Djafarian, K. (2017). *Nigella sativa* improves glucose homeostasis and serum lipids in type 2 diabetes: A systematic review and meta analysis. *Complementary Therapies in Medicine*, *35*, 6–13.

[126] Sayed-Ahmed, M. M. & Nagi, M. N. (2007). Thymoquinone supplementation prevents the development of gentamicin-induced acute renal toxicity in rats. *Clinical and Experimental Pharmacology Physiology*, *34*, 399–405.

[127] Elsherbiny, N. M. & El-Sherbiny, M. (2014). Thymoquinone attenuates Doxorubicin-induced nephrotoxicity in rats: role of Nrf2 and NOX4. *Chemico-Biological Interactions*, *223*, 102–108

[128] Awad, A. S., Al Haleem, E. N. A., El-Bakly, W. M. & Sherief, M. A. (2016). Thymoquinone alleviates nonalcoholic fatty liver disease in rats via suppression of oxidative stress, inflammation, apoptosis. *Naunyn Schmiedebergs Archives of Pharmacology*, *389*, 381–391.

[129] Evirgen, O., Gökçe, A., Ozturk, O. H., Nacar, E., Onlen, Y. & Ozer, B. (2011). Effect of thymoquinone on oxidative stress in Escherichia coli–induced pyelonephritis in rats. *Current Therapeutic Research*, *72*, 204–215.

[130] Ince, S., Kucukkurt, I., Demirel, H. H., Turkmen, R. & Sever, E. (2012). Thymoquinone attenuates cypermethrin induced oxidative stress in Swiss albino mice. *Pesticde Biochemistry and Physiology*, *104*, 229–235.

[131] Gore, P. R., Prajapati, C. P., Mahajan, U. B., Goyal, S. N., Belemkar, S., Ojha, S. & Patil, C. R. (2016). Protective effect of thymoquinone against cyclophosphamide-induced hemorrhagic cystitis through inhibiting DNA Damage and upregulation of Nrf2 expression. *International Journal of Biological Sciences*, *12*, 944-953.

[132] Ojha, S., Azimullah, S., Mohanraj, R., Sharma, C., Yasin, J., Arya, D. S. & Adem, A. (2015). Thymoquinone protects against myocardial ischemic injury by mitigating oxidative stress and inflammation. *Evidence-Based Complementary and Alternative Medicine*, Article ID 143629, 12 pages.

[133] Liu, H., Liu, H. Y., Jiang, Y. N. & Li, N. (2016). Protective effect of thymoquinone improves cardiovascular function, and attenuates oxidative stress, inflammation and apoptosis by mediating the PI3K/Akt pathway in diabetic rats. *Molecular Medicine Reports*, *13*(3), 2836-2842.

[134] Randhawa, M. A., Alghamdi, M. S. & Maulik, S. K. (2013). The effect of thymoquinone, an active component of Nigella sativa, on isoproterenol induced myocardial injury. *Pakistan Journal of Pharmaceutical Science*, *26*(6), 1215-1219.

[135] Nagi, M. N., AI-Shabanah, O., Hafez, M. M. & Sayed-Ahmed, M. M. (2011). Thymoquinone supplementation attenuates cyclophosphamide induced cardiotoxicity in rats. *J Biochem Mol Toxicol*, *25*(3), 135-142.

[136] Al-Shabanah, O. A., Badmy, O. A., Nagi, M. N., Ai-Gharably, N. M., Ai- Rikabi, A. C. & Al–Bekairi, A. M. (1998). Thymoquinone protects against doxorubicin-induced cardiotoxicity without compromising its antitumor activity. *Journal of Experimental & Clinical Cancer Research*, *17*(2), 193-198.

[137] Banerjee, S., Padhye, S., Azmi, A., Wang, Z., Philip, P., Kucuk, O., Sarkar, F. & Mohammed, R. M. (2010). Review on Molecular and Therapeutic Potential of Thymoquinone in Cancer. *Nutrition and Cancer*, *62*(7), 938–946.

[138] Banerjee, S, Kaseb, A. O., Wang, Z., Kong, D., Mohammad, M., Padhye, S., Sarkar, F. H. & Mohammad, R. M. (2009). Antitumor activity of gemcitabine and oxaliplatin is augmented by thymoquinone in pancreatic cancer. *Cancer Research*, *69*, 5575–5583.

[139] Yamasaki, L. (2003). Role of the RB tumor suppressor in cancer. *Cancer Treatment and Research*, *115*, 209–239.

[140] Alhosin, M., Abusnina, A., Achour, M., Sharif, T., Muller, C., Peluso, J., Chataigneau, T., Lugnier, C., Schini-Kerth, V. B., Bronner, C. & Fuhrmann, G. (2010) Induction of apoptosis by thymoquinone in lymphoblastic leukemia Jurkat cells is mediated by a p73-dependent pathway which targets the epigenetic integrator UHRF1. *Biochemical pharmacology*, *79*, 1251–1260.

[141] Effenberger, K., Breyer, S. & Schobert, R. Terpene conjugates of the Nigella sativa seed-oil constituent thymoquinone with enhanced efficacy in cancer cells. *Chemistry & Biodiversity*, *7* (1), 129–139.

[142] Banerjee, S., Azmi, A. S., Padhye S, Singh M. W., Baruah, J. B., Philip, P. A. & Sarkar, F. H. (2010). Structure-activity studies on therapeutic potential of thymoquinone analogs in pancreatic cancer. *Pharmaceutical Research*, *27* (6), 1146-1158.

[143] Ravindran, J., Nair, H. B., Sung, B., Prasad, S., Tekmal, R. R. & Aggarwal, B. B. Thymoquinone poly (lactide-coglycolide) nanoparticles exhibit enhanced antiproliferative, anti-inflammatory, and chemosensitization potential. *Biochemical Pharmacology*, *79*, 1640–1647.

[144] Gökce, E. C., Kahveci, R., Gökce, A., Cemil, B., Aksoy, N., Sargon, M. F., Kisa, U., Erdogan, B., Guvenc, Y., Alagoz, F. & Kahveci, O. (2016). Neuroprotective effects of thymoquinone against spinal cord ischemia reperfusion injury by attenuation of inflammation, oxidative stress, and apoptosis. *Journal of Neurosurgery, Spine*, *24*, 949–959.

[145] Abukhader, M. (2012). The effect of route of administration in thymoquinone toxicity in male and female rats. *Indian Journal of Pharmaceutical Sciences*, *74*, 195.

[146] Zafeer, M. F., Waseem, M., Chaudhary, S. & Parvez, S. (2012). Cadmium induced hepatotoxicity and its abrogation by thymoquinone. *Journal of Biochemical and Molecular Toxicology*, *26*, 199–205.

[147] Badary, O., Ai-Shabanah, O., Nagi, M., Ai-Rikabi, A. & Elmazar, M. (1999). Inhibition of benzo (a) pyrene-induced forestomach carcinogenesis in mice by thymoquinone. *European Journal of Cancer Prevention*, *8*, 435–440.

[148] Ali, B. & Blunden, G. (2003). Pharmacological and toxicological properties of *Nigella sativa. Phytotherapy Research*, *17*, 299–305.

BIOGRAPHICAL SKETCH

Sunita Singh

Affiliation: Department of Chemistry, Navyug Kanya Mahavidyalaya, University of Lucknow

Education: MSc, PhD

Research and Professional Experience:

- Worked as Research Assistant (2010-2011) on the project title: Chemistry, Antioxidant and Antimicrobial activities of Oleoresins extracted from Cardamom, Black pepper and Caraway
- Joined as JRF (UGC) for pursuing PhD (2010-2015) on the Title: "Chemistry, antioxidant, antimicrobial activities of essential oils and oleoresins of spices"
- Quantitative and qualitative analyses of essential oils and oleoresins of spices namely *Piper nigrum*, *Nigella sativa*, *Mentha longifolia*, *Anethum graveolans*, *Brassica juncea* and *Sinapis alba* were done with the help of Gas chromatography and Mass spectrometry.

- Antioxidant and antimicrobial efficacies of essential oil and oleoresins were also investigated using different techniques because nowadays most of the quality food products have replaced the use of pure spices with spice oleoresins and spice oils. Therefore, this encourages us to carry out the research on essential oils and oleoresins of spices for their possible use as food preservatives.
- It is clear that the essential oils obtained from taken spices do have the potential for actual application in real food systems. On the other hand, there is a need for a more rational and standardized approach in experimental design so as to generate meaningful data that can be compared among different research groups and easily transferred to the actual application of essential oils for the preservation of food as well as the manufacture of health oriented products.

Professional Appointments: Joined as Assistant Professor in Department of Chemistry, Navyug Kanya Mahavidyalaya, University of Lucknow in April 2019.

Honors: Qualified National Eligibility Test (Junior Research Fellow with Rank 0448/0973) and also awarded with UGC senior research fellowship (2013-2015).

Publications from the Last 3 Years:

Sunita, Singh., Das, S. S., Gurdip, Singh., Carola, Schuff., Marina, P de lampasona. & Cesar, Catalan. (2015). "*In vitro* antioxidant potentials and chemistry of essential oils and oleoresins from fresh and sun dried *Mentha longifolia* L.". *Journal of Essential Oil Research, 27*(1), 61-69. ISSN: 1041-2905 (Print), 2163-8152 (Online) Impact factor: 0.815

Sunita, Singh., Das, S. S., Gurdip, Singh., Carola, Schuff., Marina, P de lampasona. & Cesar, Catalan. (2015). "Comparative Study of

Chemistry and Antimicrobial Potentials of Dried and Fresh Mentha longifolia L. Essential oils and Oleoresins", *Journal of Coastal Life Medicine, 3* (12), 987-991. ISSN: 2309-5288 (Print), 2309-6152 (Online) Impact factor: N/A

Sunita, Singh., Das, S. S., Gurdip, Singh., Carola, Schuff., Marina, P de lampasona. & Cesar, Catalan. (2017). "Comparison of chemical composition, antioxidant and antimicrobial potentials of essential oils and oleoresins obtained from seeds and leaves of *Anethum graveolans* L", *Toxicology Open Access, 3,* 119. doi:10.4172/2476-2067.1000119. ISSN: 2476-2067.

Sunita, Singh., Das, S. S., Gurdip, Singh., Carola, Schuff., Marina, P de lampasona. & Cesar, Catalan. (2017). "Comparison of chemical composition, antioxidant and antimicrobial potentials of essential oils and oleoresins obtained from seeds of *Brassica juncea* and *Sinapis alba*." *MOJ Food Processing & Technology, 4*(4), 00100. ISSN: 2381-182X. Impact factor: N/A

In: *Nigella sativa*
Editor: Sanjin Berghuis

ISBN: 978-1-53617-538-7

Chapter 4

INFLUENCE OF PLANT DENSITY AND FERTILIZATION ON YIELD AND CRUDE PROTEIN OF *NIGELLA SATIVA* L.: AN ALTERNATIVE FORAGE AND FEED SOURCE

Ioannis E. Roussis*[1,*]*, Ioanna Kakabouki*[1]*,
***Eleni Tsiplakou*[2] *and Dimitrios Bilalis*[1]**

[1]Laboratory of Agronomy, Department of Crop Science,
School of Plant Sciences, Agricultural University of Athens,
Athens, Greece
[2]Laboratory of Nutritional Physiology and Feeding,
Department of Animal Science and Aquaculture,
School of Animal Biosciences, Agricultural University of Athens,
Athens, Greece

* Corresponding Author's E-mail: iroussis01@gmail.com.

ABSTRACT

Nigella sativa L. has been recognized as one of the most important medicinal plants in many parts of the world for centuries. It is also a valuable source of carbohydrates, proteins, essential fatty acids, vitamins, and minerals. Because of its exceptional nutritional and medicinal properties, this plant has been considered as one of the main sources of nutrition and healthcare for humans and animals. Literature survey revealed that dietary *N. sativa* seeds greatly improved animal products quality without compromising growth, productivity, as well as organoleptic quality. *N. sativa* is mainly cultivated as a seed crop, and there is no information on biomass quality and its potential for animal feeding. A field experiment was conducted to determine the influence of plant density and fertilization on seed and biomass yield and quality of *N. sativa* crop in order to define alternatives to local forage and feed sources for animals feeding in the Mediterranean region. The 2-year experiment was laid out in a split-plot design with two main plots (plant densities: 200 and 300 plants m^{-2}), four sub-plots [fertilization treatments: control (untreated), seaweed compost, farmyard manure, and inorganic fertilizer] and three replications for each treatment. The results indicated that *N. sativa* yields were affected by both plant density and fertilization. The highest seed yield (749-800 kg ha^{-1}) was found in low density plots fertilized with inorganic fertilizer, while the highest biomass yield (3755-4346 kg ha^{-1}) was observed in high density plots fertilized with inorganic fertilizer. Concerning the crude protein (CP) content, the highest content of total aboveground biomass (22.43-24.17%) was measured at 75 days after sowing and was achieved in plots with low density and inorganic fertilization. The highest seed CP content (26.94-27.45%) was also found under low density and inorganic fertilization. Ash and ether extract content were not influenced by plant densities and fertilization regimes. In addition, acid detergent fiber (ADF) and neutral detergent fiber (NDF) were not affected by plant density but there were significant differences between fertilization treatments. The highest ADF (35.5-36.8% and 19.7-20.0% for total aboveground biomass and seed, respectively) and NDF (49.2-50.1% and 30.4-31.1% for total aboveground biomass and seed, respectively) were found under inorganic fertilization at 115 DAS. As a conclusion, *N. sativa* could be successfully used as a novel feed crop.

Keywords: compost, feed additive, forage crop, inorganic fertilizer, nutritive value

INTRODUCTION

As the global population is projected to reach 9.8 billion people by 2050, and 11.2 billion in 2100, it is expected that there will be an increasing demand for food (UN 2017). It is predicted that global meat consumption is of an increase of about 73% by 2050, while at the same time, dairy production will be increased by 58% (FAO 2011). In this context, global animal production will face significant challenges over the next decades in order to meet the increasing demand for animal protein. Sustainable production, efficient use of natural resources as well as improvement of animal welfare constitute important factors to share the focus when attempting to meet this intensive demand (Ponnampalam, Holman, and Kerry 2016). Therefore, animal nutrition can play an important role in achieving the above-mentioned requirements. Consequential benefits related to increased efficiency of animal production and improved body composition are strongly associated with animal nutrition, and the maintenance of dietary nutrition during the critical stages of animal life can impact livestock production (Demeyer and Doreau 1999).

In general, nutrition can be a serious limitation to livestock production, especially when feed resources are insufficient in both quality and quantity. In the past, the increase in animal number was not always accompanied by an improved availability of feed resources and may result in reduced animal performance and health (Kaasschieter et al. 1992; Wanapat, Kang, and Polyorach 2013). In addition, the high demand for livestock products place pressure on the need for high producing animals. Insufficient feed quality and quantity impede increased animal production (Tona 2018). Consequently, the high demand for feed protein necessitates exploration to obtain alternative feed sources with a high nutrient content which are safe and can be introduced into diets of animals in order to enhance animal health and achieve more output.

Nigella sativa L. is a medicinal plant of *Ranunculaceae* family with a one-year life span which is native to southern Europe, North Africa, and South and West Asia. The plant height ranges from 40 to 90 cm, and the leaves are finely divided and linear with grayish-green color (Ahmad et al. 2013). The flowers are of pale blue and white color with 5-10 pedals. The fruit is an inflated capsule composed of 3-7 united follicles. Each capsule contains a various number of dark grey or black small sized (1-5 mg) seeds (Valadabadi and Farahani 2013).

N. sativa seeds have been used to promote health for countries, especially in the Middle East and Southest Asia. The seeds have been widely used in traditional medicine as digestive and appetite stimulant, antidiarrheal, anthelmintic, analgesic, and antibacterial agents (Saleh Al-Jassir 1992; Riaz, Syed, and Chaudhary 1996). Moreover, recent research studies have shown *N. sativa* to have antidiabetic, anticancer, anti-inflammatory, spasmolytic, bronchodilatory, hepatoprotective, gastroprotective, nephron-protective, antihypertensive, and possessing antioxidant effects (Riaz, Syed, and Chaudhary 1996; Ahmad et al. 2013; Kazemi 2014).

The seeds of *N. sativa* are composed of crude protein (CP) (19.6-22.3%), ether extract (EE) (36.0-40.4%), crude fiber (7.0-8.8%), total carbohydrates (29.3-34.5%) and ash (4.1-4.7%) on a fresh weight basis (Saleh Al-Jassir 1992). Additionally, the seeds also contain an exceptional amount of minerals, e.g., Fe, Na, Cu, Zn, P, Ca, and vitamins like ascorbic acid, niacin, thiamine, pyridoxine and folic acid (Takruri and Dameh, 1998). *N. sativa* seeds are characterized by 30-35% fixed oil and 0.5-1.5% essential oil with several uses for pharmaceutical and food industries (Ashraf, Ali, and Iqbal 2006; Roussis et al. 2017). Its essential oil consisting of p-cymene (32.05%), thujene (6.0%), pinene (1.11%) and camphene (11.0%) (Kazemi, 2014). Pharmacologically active constituents of essential oil are thymoquinone, dithymoquinone, thymohydroquinone as well as thymol (Ghosheh, Abdulghani, and Crooks 1999).

Feed intake and feed efficiency are the main factors to evaluate feed quality and animal growth performances. A number of studies have shown that the incorporation of *N. sativa* seeds in animal rations improves feed intake, digestibility coefficients and nutritive values in agricultural livestock (Nasr et al. 1996; Akhtar, Nasir, and Rehman 2003; Abo El-Nor et al. 2007; El-Ghousein 2010; Khan et al. 2012; Kumar and Patra 2017).

More specifically, an improvement in the feed conservation ratio of broilers was found when their diet supplemented with 1% (Guler et al. 2006) and 1.5% (Al-Beitawi and El-Ghousein 2008) of *N. sativa* seeds. The same was observed in broiler performance when in their diet incorporated found that broiler performance was improved by supplementing with 4% of *N. sativa* seeds Toghyani et al. (2010). Denli, Okan, and Uluocak (2004) reported that in quails, the addition of 1% of *N. sativa* seeds significantly increased egg weight, eggshell weight as well as eggshell thickness. Concerning the growth performance of ruminants, Zanouny et al. (2013) demonstrated that *N. sativa* seeds when added in sheep diet had a positive effect on growth performance. In addition, they found that the total digestible nutrients, metabolizable energy, digestible CP, and the digestibility coefficients of dry matter (DM), EE, and nitrogen-free extract, as well increased in sheep fed with the basal diet combined with 1% or 2% *N. sativa* seeds, compared with the control one.

N. sativa seed meal, the residual by-product of pressed *N. sativa* seed, can also be used safely and economically in animal feeding as a relatively good source of energy and protein (Silvia et al. 2012; Thayalini, Yaakub, and Alimon 2018). El-Bagir et al. (2010) reported that up to 15% of *N. sativa* meal resulted in an increase in total weight gain, feed conversion ratio, and feed intake in rabbits, while higher percentages of *N. sativa* meal in the feed rations can significantly decrease the growth performance of rabbit.

Forage constitutes the key component in dairy ration (Van Soest, Robertson, and Lewis 1991). Forage provides effective fiber in dairy rations where 75% of ration neutral detergent fiber should come from coarse forages (Mertens 1997). It has to be noted that inadequate levels of dietary fiber are associated with low milk fat, rumen acidosis as well as digestive inefficiency (Orskov and Rule 1990).

Since *N. sativa* is mainly cultivated as a seed crop, there is no information on biomass quality and its potential as forage. Moreover, according to several research studies, the effect of climatic change is expected to lead to large reductions in crop productivity of the Mediterranean region and a strategy to cope with the increasing demand for feed production includes the introduction of alternative crops, such as *N. sativa* characterized by beneficial properties with adequate yield and good nutritional value (Bilalis et al. 2018b). The purpose of this study was to determine the influence of plant density and fertilization on seed and biomass yield and quality of *N. sativa* crop in order to define alternatives to local forage and feed sources for animals feeding in the Mediterranean region.

Materials and Methods

Site Description and Experimental Design

A 2-year field experiment was carried out in the organic experimental field of the Agricultural University of Athens (Latitude: 37°59′ 1.70″ N, Longitude: 23°42′ 7.04″ E, Altitude: 29 m above sea level) during 2017 and 2018. The soil was a clay loam (29.8% clay, 34.3% silt and 35.9% sand) with pH (1:1 H_2O) 7.34, nitrate-nitrogen (NO_3-N) 12.4 mg kg^{-1} soil, available phosphorus (P) 13.2 mg kg^{-1} soil, available potassium (K) 201 mg kg^{-1} soil, 15.99% $CaCO_3$ and 1.82% organic matter. The site was managed according to organic agricultural guidelines (EC 834/2007). Meteorological data (mean monthly temperature and precipitation) pertaining to the growing season 2017 and 2018 were obtained from the weather station of the Agricultural University of Athens and are presented in Figure 1. Total precipitation in 2017 and 2018 (from February to May) was 160.2 and 160.8 mm, respectively. The mean temperature throughout the growing season was 16.2°C for 2017 and 17.8°C for 2018.

The experiment was set up on an area of 302 m^2 according to the split plot design with two main plots (plant densities: 200 plants m^{-2} and 300 plants m^{-2}), four sub-plots [fertilization treatments: control (untreated), seaweed compost (2000 kg ha^{-1} Posidonia 1-2% N, Compost Hellas S.A.), farmyard manure (2000 kg ha^{-1}, solid, 1.52% N), and inorganic fertilizer (300 kg ha^{-1} Enpeka 15-15-15+5 S, Compo GmbH)], and three replications for each treatment. The main plot and sub- plot sizes were 42.25 m^2 (6.5 m

× 6.5 m) and 9 m^2 (3m × 3 m), respectively. The soil was prepared by ploughing at a depth of about 0.25 m. Fertilizers were applied as basal fertilization by hand on the soil surface and then harrowed. *N. sativa* was sown by hand in rows 30 cm apart at a depth of 0.5-1 cm. Seed sowing was performed on 1st February in both years (2017 and 2018). Emergence was on 25th and 19th February in 2017 and 2018, respectively. Seedlings were thinned at the four-true leaf stage to the examined plant densities, which were 200 and 300 plants m^{-2}. Throughout the experimental periods, there was no incidence of pest or disease on *N. sativa* crop. Weeds were controlled by hand-hoeing as and when needed and before canopy closure.

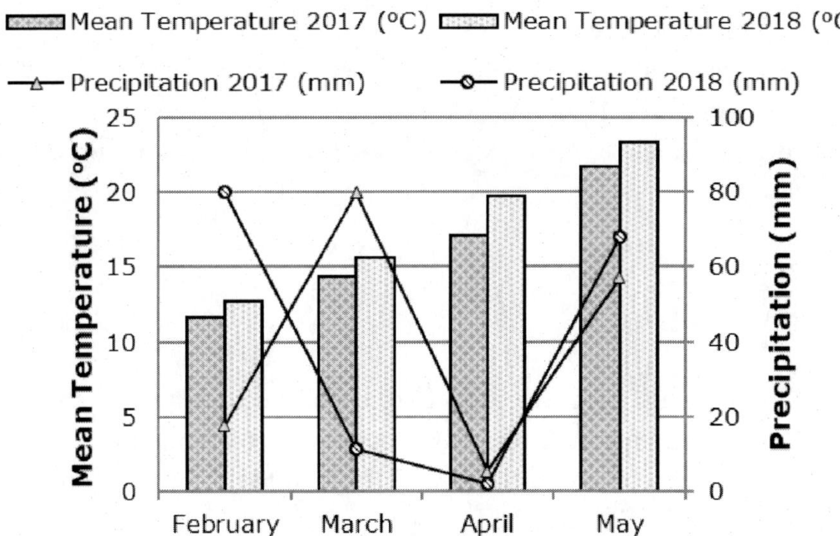

Figure 1. Weather data (mean monthly temperature and precipitation) during the experimental periods (February-May, 2017 and 2018).

Sampling, Measurements and Methods

Twenty plant samples were randomly collected from each sub-plot at 45, 60, 75, 85, 100 and 115 Days After Sowing (DAS). Above-ground dry matter was determined after drying for 48 hours at 64°C. *N. sativa* crop was harvested at physiological maturity on 3rd June 2017 (122 DAS) and

6th June 2018 (125 DAS). During harvesting, seed samples were collected from 10 randomly selected plants derived from the middle sub-plot area (1 m^2). In addition, seed yield was also determined by harvesting all the plants from the middle sub-plot area.

The plant and seed samples were ground to pass through a 1-mm screen. The samples were analyzed for ash (method 924.05), Kjeldahl nitrogen (CP; method 984.13) using a Kjeltec 8400 auto analyzer (Foss Tecator AB, Höganas, Sweden), ether extract (EE; method 920.39) and acid detergent fiber (ADF; method 973.18) according to the Association of Official Analytical Chemists (AOAC 1990). The crude protein (CP) content was calculated by multiplying the Kjeldahl nitrogen concentration by 6.25. Neutral detergent fiber (NDF; method 930.15) was determined using heat stable amylase according to the procedure of Van Soest, Robertson, and Lewis (1991). An ANKOM 200 Fiber Analyzer (ANKOM Technology Corporation, NY, USA) was used for determination of ADF and NDF.

Statistical Analysis

The experimental data were subjected to statistical analysis using the SigmaPlot 12 statistical software (Systat Software Inc., San Jose, CA, USA). The trait data produced by plant density and fertilization treatments in the two years were analyzed adopting 2 X 2 X 4 factorial design (two years; two plant density treatments and four fertilization treatments) laid out in a split-plot design with three replications. The Analysis of Variance (ANOVA) used a mixed model, with years and replications as random effects and plant density and fertilization as fixed effects. Differences between means were separated using Tukey's honestly significant difference test (Tukey's HSD). Correlation analyses were used to describe the relationships between yield components and nutritive characteristics using Pearson's correlation. All comparisons were made at the 5% level of significance ($p \leq 0.05$).

RESULTS AND DISCUSSION

Total Dry Matter (DM) Yield

The influence of plant density and fertilization on total DM yield on *N. sativa* crop are presented in Table 1. The maximum values were achieved at the end of the seed formation period (115 DAS). In the treatment with the high-density (300 plant m^{-2}), the values of total DM yield were substantially higher (3831 and 3364 kg ha^{-1} in 2017 and 2018, respectively) than the low-density treatment (2982 and 2635 kg ha^{-1} in 2017 and 2018, respectively). Seed row spacing constitutes an agronomical management strategy used by producers to optimize the husbandry of soil and plant ecosystem during the growing period with the goal of increasing the crop production. Crop row spacing affects canopy architecture, which is a distinctive characteristic that influences the utilization of light, water, and nutrients (Sharratt and McWilliams 2005). DM accumulation is directly related to the leaf area index, or the amount of radiation intercepted by the crop and optimum plant density increases total DM (Edwards, Purcell, and Vories 2005). Roussis, Kakabouki, and Bilalis (2019) demonstrated that leaf area index of *N. sativa* crop was significantly higher (2.253 m^2 m^{-2}) in the plant density of 300 plants m^{-2} than in the low density (200 plants m^{-2}) plots (1.905 m^2 m^{-2}) at flowering stage (85 DAS).

In response to fertilization, the mean values were the highest in inorganic treatment (3882 and 3412 kg ha^{-1} in 2017 and 2018, respectively) and compost (3781 and 3329 kg ha^{-1} in 2017 and 2018, respectively) followed by manure (3192 and 2802 kg ha^{-1} in 2017 and 2018, respectively) and control (2772 and 2457 kg ha^{-1} in 2017 and 2018, respectively). According to several research studies, the crops fertilized with inorganic fertilizers gave higher DM yields as these fertilizers contained soluble inorganic nitrogen, which had a quick availability for crops resulting in higher yields (Bilalis et al. 2018a; Kakabouki et al. 2018). In our study, the difference significant differences in the total DM yield among years (Table 10) will be probably related to the differences in accumulated growing degree days (ADD) between the studied cultivation periods (1417.3 and 1685.2 ADD in 2017 and 2018, respectively) (Roussis, Kakabouki, and Bilalis 2019).

Qualitative Characteristics of Total Above-Ground Biomass

The crude protein (CP) content of forages varies significantly depending on plant species, soil fertility as well as plant maturity. The CP content of alfalfa hay varies from 18 to 25% of DM (Horrocks and Valentine 1999). Moreover, the CP content and the digestibility of the forages are decreasing, while the opposite happens in their crude fiber (CF) content during their vegetative stage of growth (Rayburn 2019).

The results of our study indicated that CP content of total above-ground biomass was significantly affected by both plant density and fertilization, with the maximum value achieved between blooming and flowering stage (75 DAS) (Table 2). Concerning the effect of plant density, the CP content recorded in low-density plots (19.27% and 20.29% in 2017 and 2018, respectively) was greater than in high-density treatment (17.10% and 17.89% in 2017 and 2018, respectively). The reduction of the crude protein content at higher plant density may be due to the fact that the over-competition of crowded plant population for the available resources has made individual plants weaker, resulting in plant lodging. Over-competition and lodging of plants also make less energy available for the conversion of nitrogen into protein (Widdicombe and Thelen 2002).

Increasing levels of fertility generally increase the forage quality parameters, such as CP content. In regard to fertilization effect, the mean value was the highest in the inorganic treatement (22.13%) followed by compost (19.65%), manure (18.09%), and control (14.68%) (Table 2). The higher CP content with inorganic fertilizer may be due to the fact that higher levels of nitrogen available in plants are known to enhance nitrogen uptake, as nitrogen plays a critical role in protein synthesis (Kakabouki et al. 2018). Moreover, in several studies, it was found that plant maturity can affect the forages CP content (Xie et al. 2012; Krawutschke et al. 2013). Indeed, Throop (2005) observed that crude protein content in wheat, declined gradually with the delay of harvest time due to protein synthesis being inhibited by the weak photosynthesis at the more mature stage.

The combined analysis of variance revealed that the CP yield of total above-ground biomass was only affected by fertilization (Table 10), and the maximum values were found at the beginning of flowering stage (85 DAS). The CP yield in the control and manure treatments were quite distinct from that in the compost and inorganic fertilization treatments.

The maximum CP yield was lower in the untreated (268.7 and 371.9 kg ha^{-1} in 2017 and 2018, respectively) and significantly greater in the manure treatment (414.1 and 394.0 kg ha^{-1} in 2017 and 2018, respectively), while the greatest values were obtained in the compost (538.3 and 490.5 kg ha^{-1} in 2017 and 2018, respectively) and inorganic (609.6 and 562.3 kg ha^{-1} in 2017 and 2018, respectively) treatments. The CP yield is a function of DM yield and CP content, and as a result, the DM yield and crude CP content had a strong positive correlation with the CP yield (r=0.8920; p<0.001 and r=0.7517, p<0.001, respectively).

Ash constitutes the total mineral content in forages, which includes inorganic compounds in the plant and soil contaminants. A high ash content indicates that there is significant contamination by soil that can inflate neutral detergent fiber (NDF). The normal ash content is around 11% for alfalfa and 9% for grass forages. (Horrocks and Valentine 1999). The results of Table 4 showed that fertilization had only a significant effect on the ash content until 60 DAS (before the blooming stage).

At 115 DAS, when the highest DM yield was found, fertilization did not cause significant change in the elemental composition of *N. sativa* biomass. However, the greatest values were found in inorganic treatment (12.65% and 11.79% in 2017 and 2018, respectively), while the lowest values were obtained in manure plots (11.64% and 10.80% in 2017 and 2018, respectively). A similar trend was also observed in quinoa (*Chenopodium quinoa* Willd.) crop (Kakabouki et al. 2014). Kumar et al. (2017) reported that in maize crop, the increasing levels of inorganic fertilizer up to 150% of recommended dose (120 kg N + 60 kg P_2O_5/ha) resulted in a significant improvement of ash content (up to 8.55%) because of the higher availability of major nutrients promoted the growth and DM accumulation capacity of the plants. Contrariwise, fertilized giant reed (*Arundo donax* L.) had a lower ash concentration than the unfertilized crop (Nassi o Di Nasso, Angelini, and Bonari 2010).

CF content is considered to be a key indicator of the chemical composition when determining the energy feed value of the forage. The high CF content constitutes an indicator of low digestibility and energy feeding value of the forage (Krachunov 2007). The CF is divided into two components, neutral detergent fiber (NDF) and acid detergent fiber (ADF). Determining NDF and ADF are promoted on the basis of predicting DM intake, estimating digestible energy content, as well as ensuring adequate fiber in the diet (Beauchemin, 1996). The NDF and ADF content of the forages is of great importance especially for the lactating ruminants since the milk fat content positively relates with their percentages. The National Research Council (2001) recommends a minimum dietary NDF content of 25 to 28% and ADF content of 17 to 21% (dry matter basis) for lactating cows (NRC, 2001).

NDF measures cell wall components (hemicellulose, cellulose, and lignin), and is approximately the same as measuring the CF (Duodu and Dowell 2019). Hemicellulose and cellulose are slowly digested by rumen microbes, while lignin is indigestible. Lignin is also interlinked with other cell wall constituents, making them indigestible as well. NDF is partially digestible, depending on forage species and stage of maturity. This measure increases with advanced forage maturity and is used for the prediction of forage intake. NDF was only influenced by fertilization, and the maximum values were obtained at 115 DAS (Table 5). The greatest values were found in plants fertilized with inorganic fertilizer (50.13% and 48.03% in 2017 and 2018, respectively), and compost (47.84% and 47.87% in 2017 and 2018, respectively).

Table 1. Effect of plant density (200 and 300 plants m^{-2}) and fertilization (control, manure, compost and inorganic fertilizer) on total dry matter (DM) yield (kg ha^{-1}) by Tukey's HSD test

Fertilization	Plant Density (plants m^{-2})											
	200	300	200	300	200	300	200	300	200	300	200	300
2017	*Total DM Yield (kg ha^{-1})*											
	45 DAS		60 DAS		75 DAS		85 DAS		100 DAS		115 DAS	
Control	396	582	958	1412	1644	2348	2048	2913	2138	2984	2295	3249
Manure	423	540	1137	1445	1941	2442	2456	3070	2620	3335	2834	3550
Compost	449	598	1299	1610	2319	2848	2979	3623	3119	3901	3381	4180
Inorganic	429	564	1250	1623	2250	2913	2908	3739	3174	3973	3419	4346
$F_{plant\ density}$	14.3044** (Tukey = 70.9)		11.4091** (Tukey = 214.8)		10.7264** (Tukey = 395.9)		10.0062** (Tukey = 523.4)		9.9536** (Tukey = 573.2)		9.2617** (Tukey = 636.1)	
$F_{fertilization}$	0.2177^{ns}		1.4136^{ns}		2.5709^{ns}		3.1630* (Tukey= 788.7)		3.6832* (Tukey = 837.7)		3.4907* (Tukey = 927.1)	
$F_{plant\ density\ X\ fertilization}$	0.1424^{ns}		0.1006^{ns}		0.0739^{ns}		0.0747^{ns}		0.0119^{ns}		0.0397^{ns}	
2018												
Control	346	519	872	1258	1470	2092	1780	2539	1916	2746	2044	2868
Manure	367	476	999	1279	1744	2172	2216	2731	2322	2935	2524	3079
Compost	393	537	1079	1476	1984	2583	2548	3240	2687	3504	2904	3752
Inorganic	387	495	1125	1409	2038	2529	2649	3246	2861	3505	3068	3756
$F_{plant\ density}$	13.7043** (Tukey = 66.3)		11.8600** (Tukey = 191.56)		9.6726** (Tukey = 361.2)		8.9276** (Tukey = 476.3)		9.6351** (Tukey = 517.2)		8.6577** (Tukey = 560.9)	
$F_{fertilization}$	0.2684^{ns}		1.1031^{ns}		2.1083^{ns}		3.0144* (Tukey= 708.9)		2.9403* (Tukey = 784.0)		3.3263* (Tukey = 815.4)	
$F_{plant\ density\ X\ fertilization}$	0.1842^{ns}		0.1046^{ns}		0.0708^{ns}		0.0622^{ns}		0.0583^{ns}		0.0766^{ns}	

Different letters within a column indicate significant differences according to Tukey's HSD test ($\alpha = 0.05$). Significance levels: *$p<0.05$; **$p<0.01$; ***$p<0.001$; ns, not significant ($p>0.05$).

Table 2. Effect of plant density (200 and 300 plants m^{-2}) and fertilization (control, manure, compost and inorganic fertilizer) on crude protein (CP; %) of total above-ground biomass by Tukey's HSD test

Fertilization	Plant Density (plants m^{-2})											
	200	300	200	300	200	300	200	300	200	300	200	300
2017	*CP (%)*											
	45 DAS		60 DAS		75 DAS		85 DAS		100 DAS		115 DAS	
Control	7.07	6.32	10.77	9.63	14.98	13.42	12.68	11.37	11.36	10.15	10.55	9.96
Manure	9.59	7.49	14.61	10.61	19.53	15.63	16.54	13.59	15.44	12.37	14.32	11.48
Compost	10.42	9.56	15.61	12.92	20.12	18.44	17.05	15.62	16.32	13.96	15.13	12.99
Inorganic	10.48	9.75	15.22	14.10	22.43	20.92	19.39	16.91	16.57	14.94	14.97	13.90
$F_{plant\ density}$	3.0704^{ns}		7.7298* (Tukey = 2.172)		14.1827** (Tukey = 2.122)		14.3477** (Tukey = 1.998)		12.0380* (Tukey = 2.028)		9.8746** (Tukey =1.767)	
$F_{fertilization}$	6.3424** (Tukey = 1.863)		6.3448** (Tukey = 2.705)		30.0308*** (Tukey = 2.183)		22.8375*** (Tukey = 2.012)		13.9503*** (Tukey= 2.134)		12.7228*** (Tukey = 1.869)	
$F_{plant\ density\ X\ fertilization}$	0.2717^{ns}		0.7417^{ns}		1.0248^{ns}		0.5529^{ns}		0.4786^{ns}		0.9216^{ns}	
2018												
Control	7.53	6.69	11.92	10.71	16.01	14.29	13.56	11.89	12.14	10.85	11.47	10.64
Manure	10.25	7.80	16.04	11.66	21.00	16.21	17.62	14.29	16.43	13.01	15.27	11.98
Compost	10.63	10.06	16.21	14.43	19.98	20.07	17.09	16.72	16.35	15.12	14.99	14.25
Inorganic	11.27	10.13	17.13	15.41	24.17	20.99	20.60	17.62	17.59	15.46	16.47	14.37
$F_{plant\ density}$	3.1736^{ns}		5.4271* (Tukey = 2.235)		11.3611** (Tukey = 2.363)		8.1974* (Tukey = 2.004)		8.5262** (Tukey = 1.995)		5.8684* (Tukey = 1.486)	
$F_{fertilization}$	5.3577** (Tukey = 2.0834)		4.9039* (Tukey = 3.0860)		18.9488** (Tukey = 2.728)		13.2050*** (Tukey = 2.489)		10.2395*** (Tukey = 2.330)		6.8450** (Tukey = 2.334)	
$F_{plant\ density\ X\ fertilization}$	0.3540^{ns}		0.5356^{ns}		2.1385^{ns}		0.8511^{ns}		0.5518^{ns}		0.6970^{ns}	

Different letters within a column indicate significant differences according to Tukey's HSD test ($\alpha = 0.05$). Significance levels: *$p<0.05$; **$p<0.01$; ***$p<0.001$; ns, not significant ($p>0.05$).

Table 3. Effect of plant density (200 and 300 plants m^{-2}) and fertilization (control, manure, compost and inorganic fertilizer) on crude protein (CP) yield (kg ha^{-1}) of total above-ground biomass by Tukey's HSD test

Fertilization	Plant Density (plants m^{-2})											
	200	300	200	300	200	300	200	300	200	300	200	300
	CP Yield (kg ha^{-1})											
	45 DAS		60 DAS		75 DAS		85 DAS		100 DAS		115 DAS	
2017												
Control	28.0	36.8	103.3	136.0	246.3	315.3	260.1	331.3	243.0	303.0	242.4	323.4
Manure	40.6	42.9	166.2	162.1	378.6	384.7	405.7	422.5	404.9	419.0	406.5	414.9
Compost	46.8	57.5	202.7	211.1	466.5	530.1	507.5	571.1	509.3	549.9	511.5	548.1
Inorganic	47.6	55.2	199.6	232.1	515.9	616.0	577.6	641.6	546.3	601.2	530.3	613.0
$F_{\text{plant density}}$	1.5815^{ns}		0.5400^{ns}		1.5820^{ns}		1.0172^{ns}		0.6339^{ns}		0.9296^{ns}	
$F_{\text{fertilization}}$	2.5399* (Tukey = 16.24)		3.4898* (Tukey = 64.31)		7.0607** (Tukey = 133.19)		6.7263** (Tukey = 146.19)		6.3819** (Tukey = 143.87)		5.7584** (Tukey = 148.37)	
$F_{\text{plant density X fertilization}}$	0.0949^{ns}		0.1496^{ns}		0.1702^{ns}		0.0547^{ns}		0.0373^{ns}		0.1117^{ns}	
2018												
Control	26.1	34.7	104.1	134.7	235.4	299.1	241.8	301.9	232.6	298.1	234.9	305.5
Manure	37.7	39.6	160.3	159.1	366.1	356.7	390.0	397.9	381.9	390.9	386.5	379.1
Compost	42.0	54.2	175.3	215.9	398.8	523.8	438.0	545.9	441.6	533.1	438.3	539.4
Inorganic	45.8	50.9	199.9	222.3	501.7	533.6	555.9	586.7	516.9	552.7	520.2	550.3
$F_{\text{plant density}}$	1.4334^{ns}		0.8835^{ns}		1.3402^{ns}		0.8755^{ns}		0.9133^{ns}		0.7799^{ns}	
$F_{\text{fertilization}}$	2.1997^{ns}		2.7563* (Tukey = 67.47)		5.8859** (Tukey = 129.34)		5.4665** (Tukey = 151.76)		5.1152* (Tukey = 144.66)		4.6089* (Tukey = 151.08)	
$F_{\text{plant density X fertilization}}$	0.1431^{ns}		0.1312^{ns}		0.3861^{ns}		0.1523^{ns}		0.1151^{ns}		0.1850^{ns}	

Different letters within a column indicate significant differences according to Tukey's HSD test ($\alpha = 0.05$). Significance levels: *p<0.05; **p<0.01; ***p<0.001; ns, not significant (p>0.05).

Table 4. Effect of plant density (200 and 300 plants m^{-2}) and fertilization (control, manure, compost and inorganic fertilizer) on total ash (%) of total above-ground biomass by Tukey's HSD test

Fertilization	Plant Density (plants m^{-2})											
	200	300	200	300	200	300	200	300	200	300	200	300
	Total Ash (%)											
	45 DAS		60 DAS		75 DAS		85 DAS		100 DAS		115 DAS	
2017												
Control	7.97	7.91	10.40	10.35	10.68	10.47	10.82	10.69	11.08	11.11	11.59	11.68
Manure	7.12	7.25	9.13	9.34	10.29	10.20	10.72	10.41	11.25	10.84	11.84	11.43
Compost	8.18	8.14	10.78	10.72	10.90	10.84	11.12	11.02	11.56	11.47	12.16	12.07
Inorganic	8.65	8.36	11.50	11.09	11.62	11.18	11.84	11.40	12.27	11.83	12.88	12.43
$F_{plant\ density}$	0.0747^{ns}		0.0470^{ns}		0.4188^{ns}		0.4315^{ns}		0.3632^{ns}		0.3458^{ns}	
$F_{fertilization}$	5.9687** (Tukey = 0.822)		6.1888** (Tukey = 0.935)		2.6131^{ns}		1.5284^{ns}		1.5230^{ns}		1.7223^{ns}	
$F_{plant\ density\ X\ fertilization}$	0.1385^{ns}		0.1298^{ns}		0.1995^{ns}		0.0789^{ns}		0.0915^{ns}		0.1220^{ns}	
2018												
Control	7.44	7.52	9.31	9.37	10.04	9.96	9.95	9.97	10.62	10.89	10.91	10.90
Manure	6.60	6.78	8.15	8.37	10.19	9.31	10.41	9.83	10.63	10.14	11.15	10.45
Compost	7.65	7.81	9.10	9.98	9.95	10.50	10.15	11.55	10.60	11.00	11.03	11.46
Inorganic	8.35	7.80	10.55	9.77	11.24	10.34	11.53	10.54	11.85	11.08	12.26	11.34
$F_{plant\ density}$	0.0291^{ns}		0.1643^{ns}		0.8080^{ns}		0.7103^{ns}		0.2032^{ns}		0.8026^{ns}	
$F_{fertilization}$	9.0882*** (Tukey = 0.553)		10.8064*** (Tukey = 0.764)		1.5113^{ns}		1.9013^{ns}		1.8333^{ns}		1.8087^{ns}	
$F_{plant\ density\ X\ fertilization}$	0.7910^{ns}		2.0124^{ns}		0.9287^{ns}		0.7796^{ns}		0.7239^{ns}		0.8722^{ns}	

Different letters within a column indicate significant differences according to Tukey's HSD test ($\alpha = 0.05$). Significance levels: *$p<0.05$; **$p<0.01$; ***$p<0.001$; ns, not significant ($p>0.05$).

Table 5. Effect of plant density (200 and 300 plants m^{-2}) and fertilization (control, manure, compost and inorganic fertilizer) on Neutral Detergent Fiber (NDF; %) of total above-ground by Tukey's HSD test

Fertilization	Plant Density (plants m^{-2})											
	200	300	200	300	200	300	200	300	200	300	200	300
	NDF (%)											
	45 DAS		60 DAS		75 DAS		85 DAS		100 DAS		115 DAS	
2017												
Control	33.48	33.50	32.47	32.15	35.14	34.84	37.97	36.02	41.97	37.66	47.14	41.86
Manure	34.98	32.47	36.63	34.13	39.45	36.86	40.49	37.99	43.13	40.64	47.33	45.84
Compost	33.26	35.69	34.92	37.35	37.72	40.23	38.79	41.22	41.43	43.86	46.63	49.06
Inorganic	36.86	35.70	39.19	37.69	42.06	40.53	43.05	41.55	45.69	44.19	50.88	49.38
$F_{plant\ density}$	0.1028^{ns}		0.2144^{ns}		0.1843^{ns}		0.8033^{ns}		2.2792^{ns}		2.0444^{ns}	
$F_{fertilization}$	1.7494^{ns}		6.1452** (Tukey = 2.971)		5.4910** (Tukey = 3.208)		4.8809* (Tukey = 2.967)		4.7372* (Tukey = 3.188)		5.3156** (Tukey = 3.380)	
$F_{plant\ density\ X\ fertilization}$	1.1977^{ns}		1.0891^{ns}		0.9780^{ns}		1.2885^{ns}		2.1372^{ns}		2.3711^{ns}	
2018												
Control	31.50	32.37	31.87	32.69	33.37	33.79	36.18	35.41	40.36	37.39	47.15	41.77
Manure	33.14	32.69	34.46	32.90	36.20	34.77	39.39	36.27	42.19	39.70	47.35	45.82
Compost	31.78	34.75	32.48	36.25	34.81	38.61	37.18	40.52	40.49	42.92	46.66	49.10
Inorganic	35.02	34.43	38.06	36.19	39.93	37.32	42.18	38.87	44.75	42.58	50.95	49.43
$F_{plant\ density}$	0.4855^{ns}		0.0694^{ns}		0.0018^{ns}		0.9561^{ns}		1.6262^{ns}		1.7328^{ns}	
$F_{fertilization}$	1.3248^{ns}		3.4793* (Tukey = 3.248)		4.3841* (Tukey = 3.110)		4.0089* (Tukey = 3.209)		3.7685* (Tukey = 3.169)		4.4466* (Tukey = 3.647)	
$F_{plant\ density\ X\ fertilization}$	0.6730^{ns}		1.4315^{ns}		1.9135^{ns}		2.4493^{ns}		1.5083^{ns}		0.2347^{ns}	

Different letters within a column indicate significant differences according to Tukey's HSD test (α = 0.05). Significance levels: *p<0.05; **p<0.01; ***p<0.001; ns, not significant (p>0.05).

Table 6. Effect of plant density (200 and 300 plants m^{-2}) and fertilization (control, manure, compost and inorganic fertilizer) on Acid Detergent Fiber (ADF; %) of total above-ground biomass by Tukey's HSD test

Fertilization	Plant Density (plants m^{-2})											
	200	300	200	300	200	300	200	300	200	300	200	300
	Acid Detergent Fiber (%)											
	45 DAS		60 DAS		75 DAS		85 DAS		100 DAS		115 DAS	
2017												
Control	16.61	16.32	20.28	19.14	21.82	20.32	23.36	21.81	24.72	23.21	33.24	29.69
Manure	17.81	15.28	22.82	21.20	23.66	22.39	25.18	23.91	25.77	24.52	33.38	32.53
Compost	15.76	18.49	22.48	23.03	23.64	24.23	25.17	25.77	24.80	26.36	32.87	34.83
Inorganic	17.70	18.53	22.96	23.28	25.12	24.45	26.63	25.97	27.23	26.55	35.91	35.06
$F_{plant\ density}$	0.0621^{ns}		0.5183^{ns}		1.4827^{ns}		1.0059^{ns}		0.5907^{ns}		1.1384^{ns}	
$F_{fertilization}$	1.0446^{ns}		5.4289** (Tukey = 1.863)		7.3823** (Tukey = 1.703)		4.9301* (Tukey = 2.032)		3.9229* (Tukey= 1.825)		4.7585* (Tukey= 2.468)	
$F_{plant\ density\ X\ fertilization}$	2.1740^{ns}		0.2897^{ns}		0.6305^{ns}		0.4388^{ns}		1.3173^{ns}		2.1247^{ns}	
2018												
Control	18.45	17.37	21.59	20.08	23.69	21.26	24.34	21.89	25.24	22.78	33.99	30.38
Manure	19.52	15.07	23.47	22.09	24.72	23.28	25.40	23.93	25.90	24.48	34.15	34.67
Compost	17.13	19.36	22.60	23.93	23.76	25.12	24.42	25.78	24.94	26.32	34.79	36.70
Inorganic	19.41	19.72	24.01	24.17	26.19	25.33	26.88	26.01	27.36	26.51	36.67	36.94
$F_{plant\ density}$	0.6584^{ns}		0.2689^{ns}		2.1864^{ns}		1.8586^{ns}		2.2613^{ns}		0.0988^{ns}	
$F_{fertilization}$	1.0853^{ns}		4.1664* (Tukey = 1.963)		5.6551** (Tukey = 1.572)		4.7198* (Tukey = 1.652)		4.7350* (Tukey= 1.820)		7.4061** (Tukey= 2.363)	
$F_{plant\ density\ X\ fertilization}$	2.3393^{ns}		0.9937^{ns}		1.9849^{ns}		1.6411^{ns}		2.1281^{ns}		2.6188^{ns}	

Different letters within a column indicate significant differences according to Tukey's HSD test (α = 0.05). Significance levels: *$p<0.05$; **$p<0.01$; ***$p<0.001$; ns, not significant ($p>0.05$).

Table 7. Effect of plant density (200 and 300 plants m^{-2}) and fertilization (control, manure, compost and inorganic fertilizer) on ether extract (EE; %) of total above-ground biomass by Tukey's HSD test

Fertilization	Plant Density (plants m^{-2})											
	200	300	200	300	200	300	200	300	200	300	200	300
	EE (%)											
	45 DAS		60 DAS		75 DAS		85 DAS		100 DAS		115 DAS	
2017												
Control	1.57	1.63	2.07	2.14	3.08	3.15	2.27	2.28	2.11	2.20	2.07	2.09
Manure	1.43	1.49	1.85	1.93	2.99	2.95	2.31	2.23	2.17	2.15	2.14	2.03
Compost	1.65	1.67	2.18	2.22	3.19	3.23	2.38	2.36	2.23	2.26	2.22	2.20
Inorganic	1.71	1.72	2.28	2.29	3.28	3.30	2.52	2.43	2.33	2.34	2.41	2.30
$F_{\text{plant density}}$	0.4602^{ns}		0.3212^{ns}		0.0705^{ns}		0.3522^{ns}		0.0687^{ns}		0.2787^{ns}	
$F_{\text{fertilization}}$	3.9851* (Tukey = 0.145)		4.2842* (Tukey = 0.220)		3.2883* (Tukey = 0.201)		1.4425^{ns}		1.0047^{ns}		1.4033^{ns}	
$F_{\text{plant density X fertilization}}$	0.0425^{ns}		0.0451^{ns}		0.1061^{ns}		0.0842^{ns}		0.0722^{ns}		0.1006^{ns}	
2018												
Control	1.47	1.55	1.99	2.02	3.10	3.08	2.15	2.32	2.15	2.32	1.95	1.96
Manure	1.33	1.39	1.73	1.80	3.13	2.94	2.26	2.25	2.26	2.25	2.01	1.82
Compost	1.53	1.61	1.92	2.15	3.09	3.19	2.25	2.33	2.25	2.33	1.98	2.09
Inorganic	1.65	1.60	2.19	2.10	3.35	3.16	2.51	2.35	2.51	2.35	2.31	2.07
$F_{\text{plant density}}$	0.7517^{ns}		0.7730^{ns}		0.8086^{ns}		0.1870^{ns}		0.0103^{ns}		0.6304^{ns}	
$F_{\text{fertilization}}$	5.6340** (Tukey = 0.134)		5.0149* (Tukey = 0.208)		1.5010^{ns}		1.6726^{ns}		1.3705^{ns}		1.6148^{ns}	
$F_{\text{plant density X fertilization}}$	0.3788^{ns}		0.8486^{ns}		0.8961^{ns}		0.6563^{ns}		0.5405^{ns}		0.7930^{ns}	

Different letters within a column indicate significant differences according to Tukey's HSD test ($\alpha = 0.05$). Significance levels: *$p<0.05$; **$p<0.01$; ***$p<0.001$; ns, not significant ($p>0.05$).

The reason for the significant differences between the NDF content of the fertilization treatments may be attributed to the increasing effects of fertilization on the CF content of biomass (Messman et al. 1991). So far, it has been found that the NDF content of the forages either was not affected (Benett et al. 2008) or was reduced (Dasci and Comakli 2011) by nitrogen fertilizers. This discreapancy is associated with the season of evaluation, mainly during the vegetative growth. It is well-known that higher growing rates can result in stem accumulation, and therefore an increase in the NDF content (Sbrissia and Silva 2008).

ADF is the measure of cellulose and lignin content in the plant, and is also partially digestible. ADF has been used as a predictor of forage digestibility, since high ADF values are associated with decreased digestibility (Han et al. 2003). As with NDF, ADF was only affected by fertilization and the greatest values found at 115 DAS. Specifically, the maximum ADF value (36.15%) was achieved in the case of the inorganic fertilization, while the lowest value (31.82%) was observed in the untreated (Table 6).

In contrast to our findings, Kering et al. (2011) observed that the nitrogen fertilization caused a reduction in the ADF content of bermudagrass (*Cynodon dactylon* (L.) Pers.) forage. As previously mentioned, ADF content is generally affected by plant maturity. Indeed, higher ADF content of *N. sativa* crop was found (Table 6), when the maturity was progressive. A significant increase in the NDF, ADF, and ether extract (EE) contents of tumbleweed (*Gundelia tournefortii* L.) were reported with advancing maturity of plants (Kamalak et al. 2005).

The EE has a heterogeneous composition and is formed by lipids (galactolipids, triglycerides, and phospholipids) and all other non-polar compounds, such as phosphatides, steroids, pigments, fat-soluble vitamins, and waxes .The amount of EE in forage is generally low, with values usually less than 3% of the DM (Coleman and Henry 2002). Analysis of variance (Table 7) revealed that EE was not affected by plant densities; however, there was a tendency for higher EE content in high-density (300 plants m^{-2}) plots. The highest ether extract content was found at 75 DAS. Concerning the fertilization effect between the blooming and flowering stage (75 DAS), EE did not differ among fertilization treatments in 2018, but significant differences were found during the first year (2017), where the highest value (3.21%) was obtained in inorganic fertilized plants. Elgersma et al. (2005) found that higher nitrogen fertilization resulted in higher fatty acid concentrations in perennial ryegrass (*Lolium perenne* L.). Moreover, several studies have shown a decline IN EE content with increasing maturity (Dewhurst et al. 2001; Boufaïed et al. 2003).

Seed Yield

The research results indicated that seed yield was affected by both plant density and fertilization (Table 8). In regard to the effect of plant density, the highest seed yields (677 and 603 kg ha^{-1} in 2017 and 2018, respectively) were found in low-density (200 plants m^{-2}), while the lowest (447 and 507 kg ha^{-1} in 2017 and 2018, respectively) in high-density (300 plants m^{-2}) plots. Concerning the fertilization effect, the mean value of seed yield was the greatest in the inorganic treatment (678 and 704 kg ha^{-1} in 2017 and 2018, respectively) followed by compost (637 and 620 kg ha^{-1} in 2017 and 2018, respectively), manure (504 and 507 kg ha^{-1} in 2017 and 2018, respectively), and control (430 and 388 kg ha^{-1} in 2017 and 2018, respectively).

In the present study, the seed yield was higher in the low-plant density, with the average value being 34% higher in comparison to high-plant density. Mollafilabi et al. (2010) observed that an increase of plant density from 180 to 240 plants m^{-2}, reduced the seed yield by 38% due to intraspecific competition. Moreover, the seed yield increased with the increase of nitrogen fertilization rates. Specifically, the lowest seed yields were obtained in control and manure treatments with a significant increase in compost and inorganic fertilized plots. Several studies have proved that plants fertilized with inorganic fertilizers produce higher yields, since these fertilizers contain soluble inorganic nitrogen and other nutrients with ready availability to crops, and therefore they help crops to produce higher yields (Riahi et al. 2009; Bilalis et al. 2018a; Kakabouki et al. 2018).

Table 8. Effect of plant density (200 and 300 plants m^{-2}) and fertilization (control, manure, compost and inorganic fertilizer) on seed yield (kg ha^{-1}) by Tukey's HSD test

Fertilization	Plant Density (plants m^{-2})			
	200	300	200	300
	Seed Yield (kg ha^{-1})			
	2017		2018	
Control	515	345	405	371
Manure	614	394	586	428
Compost	780	493	670	569
Inorganic	800	555	749	658
$F_{plant\ density}$	22.2801*** (Tukey = 124.8)		4.8846* (Tukey = 88.60)	
$F_{fertilization}$	5.5472** (Tukey = 191.2)		9.8899*** (Tukey = 180.46)	
$F_{plant\ density\ X\ fertilization}$	0.2455^{ns}		0.3422^{ns}	

Different letters within a column indicate significant differences according to Tukey's HSD test (α = 0.05). Significance levels: *$p<0.05$; **$p<0.01$; ***$p<0.001$; ns, not significant ($p>0.05$).

Qualitative Characteristics of Seed

The CP content in *N. sativa* seed had negative response to plant density (Table 9). The highest CP content (23.15% and 23.27% in 2017 and 2018, respectively) was recorded when plants subjected to low-density (200 plants m^{-2}). The higher CP content in low plant density was probably due to the more availability of resources such as nutrients, water, and light, which might have promoted the synthesis of proteins. Czarnik, Jarecki, and Bobrecka-Jamro (2017) reported that in camelina (*Camelina sativa* L.), the increase in seeding rate to 400 seeds m^{-2} resulted in the reduction of the CP content in, and increased the EE content and fat yield. On the other hand, according to Bellaloui et al. (2015), an increment of plant density promoted an increase in the CP concentration of soybean seeds.

Table 9. Effect of plant density (200 and 300 plants m^{-2}) and fertilization (control, manure, compost and inorganic fertilizer) on crude protein (CP; %), crude protein (CP) yield (kg ha^{-1}), total ash (%), Acid Detergent Fiber (ADF; %), Neutral Detergent Fiber (NDF; %), and ether extract (EE; %) of seed by Tukey's HSD test

Fertilization	Plant Density (plants m^{-2})											
	200	300	200	300	200	300	200	300	200	300	200	300
	CP (%)		*CP Yield (kg ha^{-1})*		*Total Ash (%)*		*NDF (%)*		*ADF (%)*		*EE (%)*	
2017												
Control	18.01	17.53	92.5	60.4	4.24	4.25	28.76	25.69	18.49	16.37	36.89	36.88
Manure	23.47	18.78	144.1	74.7	4.63	4.14	28.89	29.31	18.57	17.97	37.12	36.44
Compost	24.17	22.14	188.7	110.3	4.41	4.73	29.42	31.01	18.28	19.24	37.00	37.43
Inorganic	26.94	25.13	220.0	141.4	5.55	4.74	31.00	31.24	19.97	19.38	38.21	37.31
$F_{plant\ density}$	8.9821** (Tukey = 2.054)		16.9338*** (Tukey = 45.41)		0.7280^{ns}		0.1081^{ns}		2.0476^{ns}		0.8005^{ns}	
$F_{fertilization}$	21.4265*** (Tukey = 2.676)		8.4351** (Tukey = 60.76)		1.8739^{ns}		7.1768** (Tukey = 2.0146)		5.1668* (Tukey = 1.360)		1.8026^{ns}	
$F_{plant\ density\ X\ fertilization}$	1.3749^{ns}		0.4970^{ns}		0.7688^{ns}		2.5653^{ns}		2.3201^{ns}		0.8726^{ns}	
2018												
Control	19.05	16.20	77.2	60.2	4.88	4.85	28.44	25.42	19.17	17.01	36.55	36.54
Manure	23.87	18.41	140.9	79.9	4.97	4.62	28.58	27.86	19.24	18.63	36.79	36.08
Compost	22.69	22.78	152.1	130.9	5.21	5.13	28.15	29.85	18.94	19.93	36.67	37.12
Inorganic	27.45	23.84	208.5	157.8	5.80	5.45	30.76	30.02	20.69	20.07	37.95	36.99
$F_{plant\ density}$	13.6126** (Tukey = 2.874)		6.9182* (Tukey = 45.40)		0.4402^{ns}		1.1243^{ns}		2.3702^{ns}		0.9835^{ns}	

Table 9. (Continued)

Fertilization	Plant Density (plants m^{-2})											
	200	300	200	300	200	300	200	300	200	300	200	300
	CP (%)		*CP Yield (kg ha^{-1})*		*Total Ash (%)*		*NDF (%)*		*ADF (%)*		*EE (%)*	
$F_{fertilization}$	17.4019*** (Tukey = 3.164)		11.5524*** (Tukey = 46.67)		1.5031^{ns}		4.8899* (Tukey = 2.099)		6.0783** (Tukey = 1.320)		1.2514^{ns}	
$F_{plant\ density\ X\ fertilization}$	2.0733^{ns}		0.5798^{ns}		0.0760^{ns}		2.1729^{ns}		2.7566^{ns}		1.0917^{ns}	

Different letters within a column indicate significant differences according to Tukey's HSD test ($\alpha = 0.05$). Significance levels: *$p<0.05$; **$p<0.01$; ***$p<0.001$; ns, not significant ($p>0.05$).

Table 10. Combined analysis of variance (*F* values) for yield and qualitative characteristics of total above-ground biomass and seed of *N. sativa*

Source of Variance	df	***Total Above-ground Biomass***						
		DM Yield (115 DAS)	*CP (75 DAS)*	*CP Yield (85 DAS)*	*Total Ash (115 DAS)*	*NDF (115 DAS)*	*ADF (115 DAS)*	*EE (75 DAS)*
Y	1	4.7580*	3.9247^{ns}	0.7103^{ns}	11.037**	0.0033^{ns}	6.3977^{ns}	4.7194*
PD	1	17.8874***	24.8784***	1.8872^{ns}	1.0752^{ns}	3.7431^{ns}	0.9810^{ns}	0.8720^{ns}
F	3	6.8013**	46.6129***	12.1241***	3.0261^{ns}	9.6627***	11.8180***	2.9408*
Y X PD	1	0.1032^{ns}	0.0675^{ns}	0.0018^{ns}	0.0270^{ns}	0.0051^{ns}	0.3113^{ns}	0.0163^{ns}
Y X F	3	0.0351^{ns}	0.0181^{ns}	0.0271^{ns}	0.0162^{ns}	0.0074^{ns}	0.2064^{ns}	0.0170^{ns}
PD X F	3	0.0845^{ns}	2.7496^{ns}	0.1646^{ns}	0.7293^{ns}	4.2991*	4.5920**	0.6552^{ns}
Y X PD X F	3	0.0275^{ns}	0.6497^{ns}	0.0456^{ns}	0.1892^{ns}	0.0037^{ns}	0.1254^{ns}	0.1712^{ns}

Source of Variance	**df**	***Seed***						
		Seed Yield	*CP*	*CP Yield*	*Total Ash*	*NDF*	*ADF*	*EE*
Y	1	0.0511^{ns}	0.1804^{ns}	0.0834^{ns}	6.2535*	2.9535^{ns}	5.7492*	2.0378^{ns}
PD	1	24.9376***	22.4799***	23.1825***	1.1389^{ns}	0.9873^{ns}	4.3986^{ns}	1.7729^{ns}
F	3	14.7505***	38.5357***	19.6173***	3.2963^{ns}	11.7838***	11.1828***	2.6259^{ns}
Y X PD	1	4.2092^{ns}	0.4097^{ns}	1.6363^{ns}	0.0094^{ns}	0.2907^{ns}	0.0011^{ns}	0.0017^{ns}
Y X F	3	0.1903^{ns}	0.0364^{ns}	0.0693^{ns}	0.0551^{ns}	0.1804^{ns}	0.0016^{ns}	0.0031^{ns}
PD X F	3	0.4160^{ns}	2.6638^{ns}	0.8079^{ns}	0.6153^{ns}	4.5863**	5.0466**	1.9505^{ns}
Y X PD X F	3	0.1607^{ns}	0.8290^{ns}	0.2608^{ns}	0.1817^{ns}	0.1344^{ns}	0.0019^{ns}	0.0016^{ns}

Y: Year, PD: Plant Density, F: Fertilization. Significance levels: *$p<0.05$; **$p<0.01$; ***$p<0.001$; ns, not significant ($p>0.05$).

In regard to fertilization, the mean value of CP was the highest in the inorganic treatement (26.04% and 25.65% in 2017 and 2018, respectively) followed by compost (23.16% and 22.74% in 2017 and 2018, respectively), and manure (21.13% and 21.14% in 2017 and 2018, respectively). These results are consistent with the findings reported by Ashraf, Ali, and Rha (2005), Shah (2008), and Aytac et al. (2017), who demonstrated that the CP of *N. sativa* seed is increased when nitrogen is applied in excess of that required to produce maximum yields.

As with seed yield and the CP content of the seed, the CP yield was also significantly affected by the different plant densities and fertilization treatments (Table 9). Specifically, the maximum CP yield values (161.3 and 144.7 kg ha^{-1} in 2017 and 2018, respectively) were achieved in the case of the low plant density (200 plants m^{-2}). In regard to fertilization, the highest values (180.7 and 183.2 kg ha^{-1} in 2017 and 2018, respectively) were recorded in inorganic treatment, while the lowest were obtained in untreated (76.5 and 68.7 kg ha^{-1} in 2017 and 2018, respectively). The seed yield and seed CP content had a strong positive correlation with seed crude protein yield (r=0.9740; p<0.001 and r=0.9124, p<0.001, respectively).

The ash content of *N. sativa* seeds was not signicantly affected by any of the plant density or fertilization treatments, although there was a tendecy for ash content to be higher with the increase of available nitrogen and decrease of plant density. During the growing seasons, the highest values (5.55% and 5.80% in 2017 and 2018, respectively) observed in plots with low plant density and inorganic fertilization. The ash content of *N. sativa* seeds was reported to vary from 3.77% (Babayan, Koottungal, and Halaby 1978) to 4.70% (Saleh Al-Jassir 1992). Bobrecka-Jamro, Jarecki, and Buczek (2017) demonstrated that high rates of nitrogen increased the CP content and decreased the content of EE, as well as, causing a significant decrease in ash and CF content in seeds of soya bean.

The statistical analysis of NDF contents demonstrated no significant differences between plant densities in their effect on the total content of lignin, hemicellulose, and cellulose in *N. sativa* seeds (Table 9). The NDF content of *N. sativa* seeds increased compared to that in the control after the application of organic and inorganic fertilizers. Specifically, the highest

NDF content was achieved in inorganic fertilization with the values being 31.12% (14% higher than control) and 30.39% (13% higher than control) in 2017 and 2018, respectively. Barker and Sawer (2005) and Kaur et al. (2017) found that NDF and ADF content in soybean seeds remained unchanged when nitrogen fertilizers were applied. An increased growth of plant was also confirmed by a significant correlation between seed yield and NDF content (r=0.4880; p<0.001).

An analysis of the ADF content in seeds showed that it only depended on the type of fertilizer used (Table 10). The greatest increase was observed after the inorganic fertilizer application. The mean value of ADF content reduced in the control (untreated), and the mean values in the separate treatments were as follows: inorganic: 30.76%, compost: 29.61%, manure: 28.66%, and control: 27.08%. A significant postive correlation was also observed between seed yield and ADF content (r=0.4725; p<0.001).

According to the results of this study (Table 9), the EE content of *N. sativa* seeds were not affected by plant density and fertilization; although the EE content was slightly higher in low-density plots fertilized with inorganic fertilizer (38.21% and 37.95% in 2017 and 2018, respectively). The EE content of *N. sativa* seeds was reported to vary from 35.49% (Babayan, Koottungal, and Halaby 1978) to 40.40% (Saleh Al-Jassir 1992). In earlier studies, Valinejad, Vaseghi, and Afzali (2013) and Ferreira et al. (2016) did not confirm a significant effect of fertilization with nitrogen on the EE content in soya bean seeds. A strong positive correlation was found between EE content and seed yield (r=0.5846; p<0.001).

CONCLUSION

The results indicated that *N. sativa* yields were affected by both plant density and fertilization. The highest seed yield was found in low density plots fertilized with inorganic fertilizer, while the highest biomass yield was observed in high density plots fertilized with inorganic fertilizer. In terms of qualitative characteristics of total above-ground biomass, the highest CP content was measured at 75 DAS and was achieved in plots with low density and inorganic fertilization. Moreover, the ADF and NDF were not affected by plant density; however there were significant differences between fertilization treatments. The highest ADF and NDF contents of biomass were found under inorganic treatment. The EE content was also influenced by fertilization and the highest value was observed between blooming and flowering stage (75 DAS). The high ADF, NDF and relatively high CP content characterized *N. sativa* biomass can meet the requirements of lactating animals, and therefore the production of *N. sativa* biomass as forage supplement could be of great importance. In regard to nutritive quality traits of *N. sativa* seeds, plant density and fertilzation had a significant effect on CP content of seed with the highest value observed in low density plots fertilized with inorganic fertilizer. The highest seed CP yield was found in high density plots fertilized with inorganic fertilizer. As with the ADF and NDF content of biomass, the fiber components (ADF, NDF) in *N. sativa* seed were also affected with fertilization and increased as the available nitrogen increased. As a conclusion, *N. sativa* could be successfully used as a novel feed crop in ruminants diets.

REFERENCES

Abo El-Nor, S. A. H., H. M. Khattab., H. A. Al-Alamy., F. A. Salem, and M. M. Abdou. 2007. "Effect of some medicinal plants seeds in the rations on the productive performance of lactating buffaloes." *International Journal of Dairy Science* 2:348–55.

Ahmad, Aftab, Asif Husain, Mohd Mujeeb, Shah Alam Khan, Abul Kalam Najmi, Nasir Ali Siddique, Zoheir A. Damanhouri, and Firoz Anwar. 2013. "A review on therapeutic potential of *Nigella sativa*: A miracle herb." *Asian Pacific Journal of Tropical Biomedicine* 3:337–52.

Akhtar, Muhammad Shoaib, Zahid Nasir, and Abdur Rehman Abid. 2003. "Effect of feeding powdered *Nigella sativa* L. seeds on poultry egg production and their suitability for human consumption." *Veterinarski Arhiv* 73:181–90.

Al-Beitawi, N. A., and S. S. El-Ghousein. 2008. "Effect of feeding different levels of *Nigella sativa* seeds (black cumin) on performance, blood constituents and carcass characteristics of broiler chicks." *International Journal of Poultry Science* 7:715–21.

AOAC. 1990. *Official Methods of Analysis, Association of Official Analytical Chemists. 15th ed.* USA: Washington.

Ashraf, M., Qasim Ali, and Zafar Iqbal. 2006. "Effect of nitrogen application rate on the content and composition of oil, essential oil and minerals in black cumin (*Nigella sativa* L.) seeds." *Journal of the Science of Food and Agriculture* 86:871–76.

Ashraf M., Q. Ali, and E. S. Rha. 2005. "The effect of applied nitrogen on the growth and nutrient concentration of Kalonji (*Nigella sativa*)." *Australian Journal of Experimental Agriculture* 45:459–63.

Aytac, Zehra, Nurdilek Gulmezoglu, Tugce Saglam, Engin Gokhan Kulan, Ugur Selengil, and Halit Levent Hosgun. 2017. "Changes in N, K, and fatty acid composition of black cumin seeds affected by nitrogen doses under supplemental potassium application." *Journal of Chemistry*, vol. 2017, Article ID 3162062, 7 pages.

Babayan, V. K., D. Koottungal, and G. A. Halaby. 1978. "Proximate-analysis, fatty acid and amino acid composition of *Nigella sativa* L. seeds." *Journal of Food Science* 43:1314-19.

Barker, Daniel W., and John E. Sawyer. 2005. "Nitrogen application to soybean at early reproductive development." *Agronomy Journal* 97:615–19.

Beauchemin, Karen A. 1996. "Using ADF and NDF in dairy cattle diet formulation-a western Canadian perspective." *Animal Feed Science Technology* 58:101–11.

Bellaloui, Nacer, H. Arnold Bruns, Hamed K. Abbas, Alemu Mengistu, Daniel K. Fisher, Krishna N. and Reddy. 2015. "Agricultural practices altered soybean seed protein, oil, fatty acids, sugars, and minerals in the Midsouth USA." *Frontiers in Plant Science* 6:31.

Bilalis, Dimitrios, Magdalini Krokida, Ioannis Roussis, Panayiota Papastylianou, Ilias Travlos, Nikolina Cheimona, and Argyro Dede. 2018a. "Effects of organic and inorganic fertilization on yield and quality of processing tomato (*Lycopersicon esculentum* Mill.)." *Folia Horticulturae* 30:321–32.

Bilalis, Dimitrios J., Ioannis Roussis, Nikolina Cheimona, Ioanna Kakabouki, and Ilias S. Travlos. 2018b. "Organic Agriculture and Innovative Feed Crops." In *Agricultural Research Updates*, vol. 23., edited by Prathamesh Gorawala, and Srushti Mandhatri, 55–100. New York: Nova Science Publishers.

Bobrecka-Jamro, Dorota, Wacław Jarecki, and Jan Buczek. 2018. "Response of soya bean to different nitrogen fertilization levels." *Journal of Elementology* 23:559–68.

Boufaïed, H., P. Y. Chouinard, G. F. Tremblay, H. V. Petit, R. Michaud, and G. Bélanger. 2003. "Fatty acids in forages. I. Factors affecting concentrations." *Canadian Journal of Animal Science* 83:501–11.

Coleman, S. W. and D. A. Henry. 2002. "Nutritive Value of Herbage." In *Sheep Nutrition*, edited by M. Freer and H. Dove, 1–26. Wallingford: CABI Pub. in association with CSIRO Pub.

Czarnik, Magdalena, Waclaw Jarecki, and Dorota Bobrecka-Jamro. 2017. "The effects of varied plant density and nitrogen fertilization on quantity and quality yield of *Camelina sativa* L." *Emirates Journal of Food and Agriculture* 29:988–93.

Dasci, Mahmut, and Binali Comakli. 2011. "Effects of fertilization on forage yield and quality in range sites with different topographic structure." *Turkish Journal of Field Crops* 16:15–22.

Demeyer, D., and M. Doreau. 1999. "Targets and procedures for altering ruminant meat and milk lipids." *Proceedings of the Nutrition Society* 58:593–607.

Denli, M., F. Okan, and A. N. Uluocak. 2004. "Effect of dietary black seed (*Nigella sativa* L.) extract supplementation on laying performance and egg quality of quail (*Coturnix cotnurnix* japonica)." *Journal of Applied Animal Research* 26:73–6.

Dewhurst, R. J., N. D. Scollan, S. J. Younell, J. K. S. Tweed, and M. O. Humphreys. 2001. "Influence of species, cutting date and cutting interval on the fatty acid composition of grasses." *Grass and Forage Science* 56:68–74.

Duodu, Kwaku G., and Floyd E. Dowell. 2019. "Sorghum and Millets: Quality Management Systems." In *Sorghum and Millets: Chemistry, Technology and Nutritional Attributes, Second Edition*, edited by John R. N. Taylor, and Kwaku G. Duodu, K.G, 421–40. Cambridge: Woodhead Publishing.

EC 834/2007. 2007. *(EC) No. 834/2007 of 28 June 2007 on Organic Production and Labeling of Organic Products and Repealing Regulation (EEC) No. 2092/91.* Europe: The Council of the European Union.

Edwards, Jeffrey T., Larry C. Purcell, and Earl D. Vories. 2005. "Light interception and yield potential of short-season maize (*Zea mays* L.) hybrids in the Midsouth." *Agronomy Journal* 97:225–34.

El-Bagir, Nabiela M., Imtithal T. O. Farah, Ahmed Alhaidary, Hasab E. Mohamed, and Anton C. Beynen. 2010. "Clinical laboratory serum values in rabbits fed diets containing black cumin seeds." *Journal of Animal and Veterinary Advances* 9:2532–36.

El-Ghousein, Safaa S. 2010. "Effects of some medicinal plants as feed additives on lactating awassi ewe performance, milk composition, lamb growth and relevant blood items." *Egyptian Journal of Animal Production* 47:37–49.

Elgersma, A., P. Maudet, I. M. Witkowska, and A. C. Wever. 2005. "Effects of nitrogen fertilisation and regrowth period on fatty acid concentrations in perennial ryegrass (*Lolium perenne* L.)." Annals of Applied Biology 147:145–52.

FAO. 2011. *World Livestock 2011 - Livestock in food security.* Rome: FAO.

Ferreira André Sampaio, Alvadi Antonio Balbinot Junior, Flavia Werner, Claudemir Zucareli, Julio Cezar Franchini, Henrique Debiasi. 2016. "Plant density and mineral nitrogen fertilization influencing yield,

yield components and concentration of oil and protein in soybean grains." *Bragantia* 75:362–70.

Ghosheh, Omar A., Abdulghani A. Houdi, and Peter A. Crooks. 1999. "High performance liquid chromatography analysis of the pharmacologically active quinones and related compounds in the oil of the black seed (*Nigella sativa*)." *Journal of Pharmaceutical and Biomedical Analysis* 19:757–62.

Guler, T., B. Dalkılıç, O. N. Ertas, and M. Çiftçi. 2006. "The effect of dietary black cumin seeds (*Nigella sativa* L.) on the performance of broilers." Asian- *Australasian Journal of Animal Sciences* 19:425–30.

Han, F., S. E. Ullrich, I. Romagosa, J. A. Clancy, J. A. Froseth and D. M. Wesenberg. 2003. "Quantitative genetic analysis of acid detergent fibre content in barley grain." *Journal of Cereal Science* 38:167–72.

Horrocks, Rodney, D., and John F. Valentine. 1999. *Harvested forages*. San Diego: Academic Press.

Kaasschieter, G. A., R. de Jong, J. B. Schiere, and D. Zwart. 1992. "Towards a sustainable livestock production in developing countries and the importance of animal health strategy therein." *Veterinary Quarterly* 14:66–75.

Kakabouki, I., D. Bilalis, A. Karkanis, G. Zervas, E. Tsiplakou, and D. Hela. 2014. "Effects of fertilization and tillage system on growth and crude protein content of quinoa (*Chenopodium quinoa* Willd.): An alternative forage crop." *Emirates Journal of Food and Agriculture* 26:18–24.

Kakabouki, Ioanna P., Dimitra Hela, Ioannis Roussis, Panagiota Papastylianou, Adriana F. Sestras, and Dimitrios J. Bilalis. 2018. "Influence of fertilization and soil tillage on nitrogen uptake and utilization efficiency of quinoa crop (*Chenopodium quinoa* Willd.)." *Journal of Soil Science and Plant Nutrition* 18:220–35.

Kamalak, Adem, Onder Canbolat, Yavuz Gurbuz, Adem Erol, and Osman Ozay. 2005. "Effect of maturity stage on chemical composition, in vitro and in situ dry matter degradation of tumbleweed hay (*Gundelia tournefortii* L.)." *Small Ruminant Research* 58:149–56.

Kaur, Gurpreet, William R. Serson, John M. Orlowski, Justin M. McCoy, Bobby R. Golden, and Nacer Bellaloui 2017. "Nitrogen Sources and Rates Affect Soybean Seed Composition in Mississippi." *Agronomy* 7(4):77.

Kazemi, Mohsen. 2014. "Phytochemical composition, antioxidant, anti-inflammatory and antimicrobial activity of *Nigella sativa* L. essential oil." *Journal of Essential Oil Bearing Plants* 17:1002–11.

Kering, Maru K., John A. Guretzky, Eddie Funderburg, and Jagadeesh Mosali. 2011. "Effect of nitrogen fertilizer rate and harvest season on forage yield, quality, and macronutrient concentrations in Midland Bermuda grass." *Communications in Soil Science and Plant Analysis* 42:1958–71.

Khan, Sohail H., Jahanzeb Ansari, Ahsan u. Haq, and Ghulam Abbas. 2012. "Black cumin seeds as phytogenic product in broiler diets and its effects on performance, blood constituents, immunity and caecal microbial population." *Italian Journal of Animal Science* 11:438–44.

Krachunov, I. 2007. "Estimation of energy feeding value of forages for ruminants II. Energy prediction through crude fiber content." *Journal of Mountain Agriculture on the Balkans* 10:122–34.

Krawutschke, M., J. Kleen, N. Weiher, R. Loges, F. Taube, and M. Gierus. 2013. "Changes in crude protein fractions of forage legumes during the spring growth and summer regrowth period." *Journal of Agricultural Science* 151:72– 90.

Kumar, P., and A. K. Patra. 2017. "Beneficial uses of black cumin (*Nigella sativa* L.) seeds as a feed additive in poultry nutrition." *World's Poultry Science Journal* 73:872–85.

Mertens, D. R. 1997. "Creating a system for meeting the fiber requirements of dairy cow." *Journal of Dairy Science* 80:1463–81.

Messman, M. A., W. P. Weiss, and D. O. Erickson. 1991. "Effects of nitrogen fertilization and maturity of bromegrass on in situ ruminal digestion kinetics of fiber." *Journal of Animal Science* 69:1151–61.

Nassi o Di Nasso, N., L. G. Angelini, and E. Bonari. 2010. "Influence of fertilisation and harvest time on fuel quality of giant reed (*Arundo donax* L.) in Central Italy." *European Journal of Agronomy* 32:219–27.

Nasr, A. S., A. I. Attia, A. A. Rashwan, and A. M. M. Abdine. 1996. "Growth performance of New Zealand White rabbits as affected by partial replacement of diet with *Nigella sativa* or soybean meals." *Egyptian Journal of Rabbits Sciences* 6:129–41.

NRC. 2001. *Nutrient Requirements of Dairy Cattle. 7th edition.* Washington: National Academies Press.

Orskov, E. R. and M, Ryle. 1990. *Energy nutrition in ruminants.* New York: Elsevier press.

Ponnampalam, E. N., B. W. B. Holman, and J. P. Kerry. 2016. "Impact of animal nutrition on muscle composition and meat quality." In *Meat Quality: Genetic and Environmental Factors*, edited by Wieslaw Przybylski, and David Hopkins, 101–46. Boca Raton: CRC Press.

Rayburn, E. B. 2019. "Matching plant species to your environment, weather, and climate." In *Horse Pasture Management*, edited by Paul Sharpe, 233–43. New York: Academic Press.

Riahi, Anissa, Chafik Hdider, Mustapha Sanaa, Néji Tarchoun, Mohamed Ben Kheder, Ismaïl Guezal. 2009. "Effect of conventional and organic productions systems on the yield and quality of field tomato cultivars grown in Tunisia." *Journal of the Science of Food and Agriculture* 89:2275–82.

Riaz, M., M. Syed, and F. M. Chaudhary. 1996. "Chemistry of the medicinal plants of the genus *Nigella*." *Hamdard Medicus* 39:40–5.

Roussis, Ioannis, Ioanna Kakabouki, and Dimitrios Bilalis. 2019. "Comparison of growth indices of *Nigella sativa* L. under different plant densities and fertilization." *Emirates Journal of Food and Agriculture* 31:231–47.

Roussis, Ioannis, Ilias Travlos, Dimitrios Bilalis, and Ioanna Kakabouki. 2017. "Influence of seed rate and fertilization on yield and yield components of *Nigella sativa* L. cultivated under Mediterranean semi-arid conditions." *AgroLife Scientific Journal* 6:218–23.

Saleh Al-Jassir, M. 1992. "Chemical composition and microflora of black cumin (*Nigella sativa*) seeds growing in Saudi Arabia." *Food Chemistry* 45:239–42.

Sbrissia, André Fischer, and Sila Carneiro da Silva. "Tiller size/density compensation in *Marandu palisadegrass* swards." *Revista Brasileira de Zootecnia* 37:35–47.

Shah, Shoukat Hussain. 2008. "Effects of nitrogen fertilisation on nitrate reductase activity, protein, and oil yields of *Nigella sativa* L. as affected by foliar GA_3 application." *Turkish Journal of Botany* 32:165–70.

Sharratt, Brenton S., and Danise A. McWilliams. 2005. "Microclimatic and rooting characteristics of narrow-row versus conventional-row corn." *Agronomy Journal* 97:1129–35.

Silvia, D., M. Farah Masturah, Y Tajul Aris, W. A. Wan Nadiah, and Rajeev Bhat. 2012. "The effects of different extraction temperatures of the screw press on proximate compositions, amino acid contents and mineral contents of *Nigella sativa* meal." *American Journal of Food Technology* 7:180–91.

Takruri, Hamed R. H., and Majdoleen A. F. Dameh. 1998. "Study of the nutritional value of black cumin seeds (*Nigella sativa*)." *Journal of the Science of Food and Agriculture* 76:404–10.

Thayalini, Kathiraser, Halimatun Yaakub, and Abdul Razak Alimon. 2018. "The effects of varying levels of *Nigella sativa* seed meal supplementation on nutrient digestibility and rumen fermentation characteristics in goats." *Journal of Tropical Agriculture and Food Science* 46:25–37.

Throop, Heather L. 2005. "Nitrogen deposition and herbivory affect biomass production and allocation in an annual plant." *OIKOS* 111:91–100.

Toghyani, Mehdi, Majid Toghyani, Abbasali Gheisari, Gholamreza Ghalamkari, and Mohammad Mohammadrezaei. 2010. "Growth performance, serum biochemistry and blood hematology of broiler chicks fed different levels of black seed (*Nigella sativa*) and peppermint (*Mentha piperita*)." *Livestock Science* 129:173–8.

Tona, Grace Opadoyin. 2018. "Current and Future Improvements in Livestock Nutrition and Feed Resources." In Animal Husbandry and Nutrition, edited by Banu Yücel, and Turgay Taşkin. London: Intercopen. doi: 10.5772/intechopen.73088.

UN. 2017. *World population projected to reach 9.8 billion in 2050, and 11.2 billion in 2100*. United Nations Department of Economic and Social Affairs. Last modified October 25, 2019. https://www.un.org/development/desa/en/news/population/world-population- prospects-2017.html.

Valadabadi, Sayed Alireza, and Hossein Aliabadi Farahani. 2013. "Influence of biofertilizer on essential oil, harvest index and productivity effort of black cumin (*Nigella sativa* L.)." *International Journal for Biotechnology and Molecular Research* 4:24–7.

Valinejad, Maryam, Sekineh Vaseghi, and Mehran Afzali. 2013. "Starter nitrogen fertilizer impact on soybean yield and quality." *International Journal of Engineering and Advanced Technology* 3:333–37.

Van Soest, P. J., J. B. Robertson, and B. A. Lewis. 1991. "Methods for dietary fiber, neutral detergent fiber, and nonstarch polysaccharides in relation to animal nutrition." *Journal of Dairy Science* 74:3583–97.

Wanapat, Metha, Sungchhang Kang, and Sineenart Polyorach. 2013. "Development of feeding systems and strategies of supplementation to enhance rumen fermentation and ruminant production in the tropics." *Journal of Animal Science and Biotechnology* 4:1–11.

Widdicombe, William D., and Kurt D. Thelen. 2002. "Row width and plant density effect on corn forage hybrids." *Agronomy Journal* 94:326–30.

Xie, Z. L., T. F. Zhang, X. Z. Chen, G. D. Li, and J. G. Zhang 2012. "Effects of maturity stages on the nutritive composition and silage quality of whole crop wheat." *Asian-Australasian Journal of Animal Sciences* 25:1374–80.

Zanouny, A. I., A. K. I. Abd Elmoty, M. T. Sallam, M. A. A. El-Barody, and A. A. Abd El-Hakeam. 2013. "Effect of *Nigella sativa* seed supplementation on nutritive values and growth performance of Ossimi sheep." *Egyptian Journal of Sheep and Goat Sciences* 8:57–63.

INDEX

A

F

G

H

I

K

L

M

N

O

P

Q

R

S

T

V

W

Y

Z